Industrial Development and Eco-Tourisms

"The ambivalent relations between fossil fuel extraction and tourism are a particularly apposite research topic in the North Atlantic, where global climate change can be seen at work in often poignant ways. The authors develop a relational political ecology approach which, combined with an 'omnivorous methodology', lends itself well to a nuanced analysis of the oil-tourism interface. This book is a valuable addition to environmental social science scholarship and essential reading for those interested in the complexities of a rapidly changing region."

—Karl Benediktsson, *Professor of Human Geography, University of Iceland, Iceland*

"Around the world, the same landscapes that harbor mineral riches are often tourist attractions, setting up tough policy dilemmas. The authors wisely build a framework for understanding and assessing these policies through the concept of 'social-ecological wellbeing,' applied to fascinating cases on the coasts on both sides of the North Atlantic. Anyone interested in the future of the planet will learn a lot from this book of political ecology."

—James M. Jasper, *Author of* The Art of Moral Protest

"Eco-tourism and extractive industries may seem unlikely bedfellows but the reality is that, in many parts of the world, governments pursue both. By examining the politics and cultures of oil and tourism in the North Atlantic, Stoddart, Mattoni and McLevey shed new light on how contradictions and conflicts between the two development paths are managed and, critically, where opportunities to promote more positive social and ecological futures lie. I thoroughly recommend it."

—Stewart Lockie, *Distinguished Professor of Sociology, James Cook University, Australia*

"In this sweeping five-nation comparison, the authors untangle the complicated interplay of offshore oil extraction and nature-based tourism for the coastal cultures and political interests of the North Atlantic region. Through the adoption of an "omnivorous" methodological approach, the authors identify how these two industries have co-existed in isolation, cooperation, and conflict. From their findings, they offer a compelling argument for the consistent inclusion of environmental groups in deliberations on these economic activities at sea, ideally for the enhanced wellbeing of North Atlantic communities and the marine environment. Their assessment is particularly salient given climate change and the future significance of the Arctic in both oil exploration and tourism."

—Patricia Widener, *Florida Atlantic University, USA*

Mark C. J. Stoddart · Alice Mattoni · John McLevey

Industrial Development and Eco-Tourisms

Can Oil Extraction and Nature Conservation Co-Exist?

palgrave macmillan

Mark C. J. Stoddart
Department of Sociology
Memorial University of Newfoundland
St John's, NL, Canada

John McLevey
Department of Knowledge Integration
University of Waterloo
Waterloo, ON, Canada

Alice Mattoni
Department of Political and Social Sciences
University of Bologna
Bologna, Italy

ISBN 978-3-030-55943-4 ISBN 978-3-030-55944-1 (eBook)
https://doi.org/10.1007/978-3-030-55944-1

Cover illustration: Daniel Osterkamp/gettyimages

This Palgrave Macmillan imprint is published by the registered company Springer Nature Switzerland AG
The registered company address is: Gewerbestrasse 11, 6330 Cham, Switzerland

Dedicated to Kumi Stoddart, from Mark.

Dedicated to Domenico, Martino and Agata, from Alice.

Dedicated to Allyson Stokes, from John.

Preface

This project developed as part of my (Stoddart's) ongoing research journey, which has looked at extractive and attractive modes of development since 2002. As an MA student at the University of Victoria, in British Columbia, Canada, I started my research career by looking at media coverage of forestry policy debate in British Columbia. This research examined how the *Vancouver Sun*, the flagship paper for the province, covered four episodes of forest policy contention under both New Democratic Party (NDP) and Liberal governments. This took place following a decade of intense conflict between environmental movements, First Nations groups, and the forestry sector in British Columbia over industrial forestry practices and the protection of remaining old growth forests. As such, a focus on media coverage of resource extraction-oriented policy debate was particularly timely. Among the major themes to emerge from this research is that news coverage was dominated by actors from government, the environmental movement, and the forest industry, while voices from First Nations communities and organizations, as well as labour union voices, were more peripheral. At the same time, media coverage largely replicated economic versus environmental protection frames, providing less visibility for social movement, First Nations, or community discourse that promoted alternative forms of forestry or land tenure.

The shift to my PhD research brought a change in focus from forestry practices to attractive development through a focus on the eco-politics of skiing in British Columbia. This shift in focus was inspired by the political and public discourse around environmental issues and sustainability in British Columbia at the time. While the forest industry was the most prominent target of British Columbia environmental movements, attractive economies—such as tourism, outdoor recreation, and the burgeoning British Columbia film industry—were lauded by many environmentalists and by some policymakers as sustainable alternatives to forestry and other forms of extractive development. Likewise, while several others were examining the social dimensions of resource extraction, far fewer were looking at tourism and outdoor recreation from an environmental sociological perspective. My first book, *Making Meaning out of Mountains: The Political Ecology of Skiing* (UBC Press, 2012), attempted to create a more nuanced and complex narrative of skiing by looking at how skiing is understood by the ski industry, skiers themselves, and social movements. The book shows how the ski industry works to represent itself as a pro-sustainable steward of mountain environments, which are often depicted through the imagery of uninhabited wilderness. By contrast, skiers value their recreational interactions with mountain environments, and the ability to be in these natural spaces is a large part of the draw of the sport. At the same time, skiers reflect critically on the environmental shortcomings of the sport, such as the energy and water footprints required to run chairlifts or snowmaking, the deforestation required to maintain ski runs, or the broader carbon footprint of travelling by car or airplane from cities and towns to ski hills. Finally, the pro-environmental stance of the ski industry is occasionally destabilized by environmental and First Nations groups that question the legitimacy of new ski development and expansion. For environmental organizations, skiing expansion is often framed as a threat to wildlife habitat and wilderness values. For several First Nations groups, skiing expansion becomes a site of conflict when it infringes onto unceded territory.

My focus on attractive development continued as I relocated from British Columbia to Atlantic Canada. On arriving at Memorial University in 2010, I began to look at nature-based tourism and changing social-ecological relationships with coastal environments in Newfoundland and Labrador. This province is historically dependent on the cod fishery, which suffered a major ecological collapse in the early 1990s. The 1992 cod fishing moratorium, brought in by the federal government, had massive economic and social effects across the province, leading to significant out-migration, especially from rural coastal communities. In the wake of the cod moratorium, the oil and gas industry emerged as the major economic driver of the province. However, tourism development also received a great deal of attention as a possible driver of economic renewal and diversification, especially for rural communities. By the 2010s, the tourism sector had become a significant presence in the provincial political economy, though still secondary to the oil and gas sector. Over the past several years, I engaged in a series of projects on tourism development in Newfoundland and Labrador as a force for reconfiguring social-ecological interactions with coastal environments.

This book marks an evolution of this research journey in two important ways. First, my previous projects have focused on either resource extraction (i.e. forestry) or attractive development (i.e. skiing, nature-based tourism), but rarely engaged in a comparative analysis of these different development pathways. As such, the present analysis is more holistic in focus than previous projects in examining how coastal societies navigate the benefits and challenges of both extractive and attractive development. Second, my previous projects have tended towards a regional focus, with data collection and analysis grounded in either British Columbia or Atlantic Canada. The present analysis, therefore, also represents a departure as it adopts a comparative scope on research sites across the broader North Atlantic region, including Denmark, Iceland, Newfoundland and Labrador, Norway, and Scotland. Both shifts provide new insights into the ways in which societies may better navigate the points of contact, tension, and conflict between resource extraction and attractive forms of development.

St. John's, Canada — Mark C. J. Stoddart
Bagno a Ripoli, Italy — Alice Mattoni
Waterloo, Canada — John McLevey

Acknowledgements

This project evolved with the invaluable support and input of many people. We thank the following research collaborators: Karl Benediktsson, Arn Keeling, Berit Kristoffersen, Patrick McCurdy, Kari Norgaard, Howard Ramos, Natalie Slawinski, and Julie Uldam. We also thank the following for their research assistance: Cole Atlin, Heather Austin, Quinn Burt, Gary Catano, Cory Collins, Gillian Davis, Paula Graham, Elahe Nezhadhossein, Mihai Sarbu, Megan Stewart, and Yixi Yang. For conversations that yielded essential insight we thank: Mette Blæsbjerg, Brigt Dale, Guðrún Þóra Gunnarsdóttir, Óli Halldórsson, Auður Ingólfsdóttir, James Jasper, Marianthi Leon, Marianne Helene Rasmussen, Liam Swiss, and Guðrún Rósa Þórsteinsdóttir.

Work in progress was presented at several conference and workshop venues: 2019 Canadian Rural Revitalization Foundation—North Atlantic Forum (St. John's, Canada); 2019 Canadian Sociological Association annual conference (Vancouver, Canada); 2019 Sunbelt Social Networks Conference of the International Network for Social Network

Analysis (Montreal, Canada); 2018 Arctic Circle Assembly (Reykjavik, Iceland); 2018 Climate Change and Energy Futures Workshop (St. John's, Canada); 2018 International Sociological Association World Congress of Sociology (Toronto, Canada); 2018 International Summer School of Political Ecology (Ljubljana, Slovenia); 2017 Canadian Sociological Association annual conference (Toronto, Canada); 2017 Korean Association for Environmental Sociology International Conference (Gumi, Korea); 2017 North Atlantic Forum (Bø in Telemark, Norway); and 2016 Challenges and Opportunities for Governance of Socio-Ecological Systems in Comparative Perspective (Waterloo, Canada). Departmental talks based on this work were given at COSMOS (Centre on Social Movement Studies), Scuola Normale Superiore (Florence, Italy); Department of Geography, Memorial University (St. John's, Canada); Department of Sociology, University of British Columbia (Vancouver, Canada); Department of Sociology, University of Edinburgh (Edinburgh, UK); and Global Fridays, University of Northern British Columbia (Prince George, Canada). Several ideas presented in the Epilogue on Methodology were developed through a workshop at the University of Luxembourg. In the context of these conference and workshop presentations, we thank the following colleagues for their insight and support: Marie Louise Aastrup, Nathan Andrews, Robert Bartłomiejski, Marie-Lise Drapeau-Bisson, Shin-Ock Chang, Julia Christensen, Douglas House, Shu-Fen Kao, James Kennedy, Arkadiusz Kolodziej, Janike Larsen, Barbara Neis, Susan O'Donnell, Georgia Piggot, Becki Ross, David B. Tindall, Catherine Mei Ling Wong, and Tuomas Ylä-Antilla.

Work on this book was supported by visiting scholar positions for Stoddart and McLevey at the Scuola Normale Superiore (Florence, Italy). Funding for this project was provided by an Insight Grant from the Social Sciences and Humanities Research Council of Canada (SSHRC).

Contents

About the Authors

Mark C. J. Stoddart is Professor in the Department of Sociology at Memorial University, with research interests in environmental sociology, social movements, and communications and culture. Professor Stoddart's previous book, *Making Meaning out of Mountains: The Political Ecology of Skiing*, was published in 2012 by UBC Press. He has been the principle investigator or co-investigator on several research grants from the Social Sciences and Humanities Research Council of Canada. His work has appeared in a wide range of international journals focused on environmental social science, social movement studies, and sociology, including: *Global Environmental Change*, *Organization & Environment*, *Energy Research & Social Science*, *Environmental Politics*, *Journal of Sustainable Tourism*, *Environmental Communication*, *Society & Natural Resources*, *Social Movement Studies*, *Acta Sociologica*, and *Environmental Sociology*.

Alice Mattoni is an Associate Professor at the Department of Political and Social Sciences at the University of Bologna. Her monograph *Media Practices and Protest Politics: How Precarious Workers Mobilise* was published by Ashgate in 2012 and reprinted by Routledge in 2016. Her work has appeared in *Social Movement Studies*, *Communication Theory*, *Information Communication & Society*, *Triple-C*, *Portal*, *Cultural Anthropology Online*, *Sociology Compass*, *Feminist Review*, *Working USA*, and *Participation and Conflict*. She co-edited the following volumes: *Spreading Protests: Social Movements in Times of Crisis* (ECPR Press, 2014), *Mediation and Protest Movements* (Intellect, 2013), and *Advances in the Visual Analysis of Social Movements* (Emerald, 2013).

John McLevey is an Associate Professor in the Department of Knowledge Integration, with appointments to Sociology & Legal Studies and Geography & Environmental Management. His research interests are in computational social science (especially network analysis and natural language processing); democracy and the politics of information, science, and knowledge; cognitive social science; and environmental sociology. He is the author of *Doing Computational Social Science* and *Face-to-Face: Communication and the Liquidity of Knowledge* (with Collins, Evans, Innes, Mason-Wilkes, and Kennedy). His article publications have appeared in journals such as *The Journal of Informetrics*, *Social Studies of Science*, and the *Canadian Review of Sociology*. He is the PI of Netlab, funded by an Early Researcher Award from the Ontario Ministry of Research and Innovation and several research grants from the Social Science and Humanities Research Council of Canada (SSHRC).

List of Figures

List of Tables

1

Introduction: Contact Points Between Offshore Oil and Nature-Based Tourism

Around the world, societies are caught between the desire for social and economic development and a plethora of ecological crises that include global climate change, biodiversity loss and species extinction, and declining ocean health. Social and economic wellbeing traditionally relies on various forms of extractive development, where natural resources like fish, forest products, fossil fuels, or minerals are removed and sold on regional and global markets. Increasingly, however, social, and economic development is also built on knowledge-based and cultural industries that sell services and experiences. Tourism, leisure, and recreation are important forms of what political scientist Timothy Luke calls "attractive development," or what geographer Nigel Thrift calls the "experience economy" (Luke 2002; Thrift 2000). In attractive development or an experience economy, communities and landscapes are valued as sites of leisure and tourism experience rather than as natural resource pools.

When pursued as parallel development paths, extractive and attractive development do not always go easily together. For example, conflicts can emerge when oil, forestry, or mineral extraction infringes on well-used and highly valued leisure or tourism landscapes (Gould 2017; Reichwein

M. C. J. Stoddart et al., *Industrial Development and Eco-Tourisms*,
https://doi.org/10.1007/978-3-030-55944-1_1

2014; Shaw and Magnusson 2002; Widener 2011). This book examines how societies in the North Atlantic navigate the relationships and tensions between extractive and attractive development.

Throughout this book, we use the language of social-ecological wellbeing, which we prefer to "sustainability." Sustainability is well-used in policy and business spheres to talk about the need to balance long-term environmental integrity with economic and social development goals. However, sustainable development is a malleable term, to the point where it often loses meaning. Scholars across a range of disciplines are searching for more satisfactory ways to think about building better social-ecological relationships. This includes the framework of resilience theory (Berkes and Ross 2013), as well as recent work, primarily developed in Latin America, on the notion of *buen vivir* (Gudynas 2011). This also includes the emerging dialogue around planetary health, which builds on the UN Sustainable Development Goals to conceptualize the multi-directional relationships between natural systems, social systems, and community health (Whitmee et al. 2015).

We prefer the language of social-ecological wellbeing for thinking about links between ecological, cultural, political, and economic dimensions of individual, community, and ecosystem health. As an alternative to sustainable development, our emphasis on social-ecological wellbeing aligns with trends towards resilience theory, *buen vivir*, or planetary health. However, the framework of social-ecological wellbeing responds to criticisms of resilience theory that it is insufficiently attentive to social dynamics, and especially power dynamics, in social-ecological systems (Davidson 2010).

Our guiding questions ask how North Atlantic societies navigate offshore oil and nature-oriented tourism (or eco-tourism) as development pathways, and whether they do so in ways that build social-ecological wellbeing. To answer our high-level questions, the substantive chapters of this book address a series of empirical questions: How are tourism and oil development models *evaluated* across societies, and what do these evaluations reveal about consequential differences in what is valued or devalued, and why? How and to what extent does the oil-tourism interface feature in formal governance? To what extent to social movements and civil society organizations intervene in and shape the oil-tourism interface?

Coastal societies around the North Atlantic navigate relationships between fossil fuel development, nature-based tourism, and a range of other coastal activities, including fisheries and shipping. We focus primarily on the intersections of oil and tourism, though we periodically touch on other sectors, including fisheries as a crucial third sector across the region. The oil-tourism interface is our focus because it provides a valuable entry point for understanding the cultures, politics, and economics of eco-tourism and industrial coastal activities. In answering our research questions, we provide insight into the relationships and tensions between extractive and attractive development more broadly.

The Oil-Tourism Interface

Our central concept of the oil-tourism interface describes a set of direct and indirect contact points that involve conflict and complementarity across these two sectors. Direct contact points include conflict over new oil exploration or extraction, such as the controversies in Ecuador examined by Widener (2011) and in Belize examined by Gould (2017). The threat of resource extraction can also create an interest in tourism to threatened environments that visitors should see before they are transformed. During the 1990s, for example, conflict over old-growth forestry on the west coast of Vancouver Island also worked as an "envirotisement" for tourism to contested areas and boosted the tourism economy of the region (Luke 2002). Conflict over oil development may play a similar role, as Widener (2009) demonstrates in her analysis of conflict over a pipeline project in Ecuador, where the threat posed by the pipeline also drew tourists to threatened areas. The energy sector itself may also serve as a tourism attractor, such as tours of oil facilities, or oil-oriented museums or museum displays, which are found in cities like St. John's, Canada, Esbjerg, Denmark; Stavanger, Norway; or Aberdeen, Scotland. Similarly, geothermal power facilities in Iceland are also signposted and interpreted for visitors as tourism sites. Finally, the energy sector provides resources, including funding, that supports tourism development, as in the case of Shetland Islands, Scotland (Jennings 2015). Similarly, oil companies often support cultural institutions or mega-sporting events like the Olympics that are linked to tourism, as Uldam (2018) notes in her analysis of contention between BP and social movements in the United Kingdom.

Beyond these direct contact points, there is a range of indirect contact points that the oil-tourism interface concept brings into focus. Infrastructure development for the energy sector, including roads, airports, hotels, and restaurants, may subsequently be used to support tourism development. Tourism mobilities—including airplane, car, tour bus, ferry, or cruise ship travel also rely on fossil fuel extraction and consumption. Oil-fuelled transportation systems form the skeleton of tourism mobility networks (Cohen et al. 2011; Luzecka 2016; Urry 2013). Finally, tourism environments are impacted by climate change, which is driven to a large degree by fossil fuel-based industries (Hall 2010). In the coastal societies of the North Atlantic, climate change impacts include sea level rise, increasingly severe storm seasons, changes to seasonal weather patterns, shifting wildlife habitat and migration patterns, as well as the transformation of Arctic and near-Arctic ecosystems.

These contact points are an ideal type, intended to clarify salient dimensions of the concept. The specific forms these contact points take differ across host societies. These contact points may involve collaboration or conflict across sectors, as well as with national, regional, or local governments. These contact points may also involve mobilization, opposition, or collaboration with environmental movements or other civil society groups. Our purpose in outlining this range of contact points—and the purpose of the book—is to highlight how these forms of development, which are often understood as siloed off from each other, are intertwined in a variety of ways that impact the social-cultural, environmental, and economic wellbeing of host societies, and which therefore deserve closer scrutiny.

Our concept of the oil-tourism interface highlights links between networks of oil extraction and distribution, and "tourism mobilities" (Sheller and Urry 2004). Tourism mobility is part of the shift towards "cosmopolitanism" as a way of engaging the world, which is characterized by "a kind of *connoisseurship*, of places, people, and cultures" that relies on "extensive mobility" (Szerszynski and Urry 2006, p. 114, italics in original). Eco-tourism is a form of travel where nature is the central tourism attractor. Eco-tourist travel is presumed to cultivate environmental awareness and provide a rationale for environmental protection (Luke 2002; Snyman and Bricker 2019; Urry and Larsen 2011). Often, eco-tourism emerges because host regions wish to diversify from economic dependence on resource extraction, or to recover from economic or ecological crises in extractive economies (George et al. 2009; Reed and Gill 1997). For example, Iceland has seen a shift from whaling to an increasing emphasis on whale watching tours as a core tourism attractor (Cunningham et al. 2012). As whale watching takes on greater economic and social importance, its image as a more sustainable form of economic development has been mobilized in opposition to the whaling industry.

However, as Waitt and Cook's (2007) research on kayaking in Thailand illustrates, environmental awareness raised through tourism is often limited to a specific experience and may not translate to broader commitments to environmentalism. Similarly, Gould's (1999) research on Ecuador and Belize shows that tourist operations that position themselves as environmentally responsible often restrict community access to resources and produce large amounts of waste. Laudati's (2010) analysis of gorilla-oriented tourism in Bwindi Impenetrable National Park, Uganda, similarly finds that while eco-tourism promises to promote ecological and social wellbeing, in this case it excludes local residents from access to land and resources, but provides few community benefits in return. Furthermore, while increased mobility cultivates a more cosmopolitan worldview, the cost is that it encourages an abstract view of the world that is detached from commitments to local places and communities (Szerszynski and Urry 2006). This research shows a discursive alignment between environmentalism and eco-tourism, but this alignment is often shallow and economically strategic.

Widener's (2009) research on Ecuador and the Philippines further highlights the less obvious connections between oil development and tourism. The threat of oil development can draw the attention of potential tourists to endangered environments that should be seen before they are transformed by oil extraction. Oil extraction can also serve as an incentive for "dark tourism" to places impacted by disasters like the Exxon Valdez wreck in Alaska (Widener 2009). Widener sums up the multifaceted relationship between oil and tourism by noting that oil development "may attract media attention to the area; inspire community challenges to the oil industry or oil impact; increase funding for tourism infrastructure; and reinforce … commitments to protect, build, or expand a tourism destination as an alternative to oil impacts or the oil industry" (Widener 2009, p. 267). The idea that oil extraction supports the development of local tourism infrastructure is reinforced by Jennings' (2015) analysis of the Shetland Islands. Oil revenues "brought independence of action" for the region and were important for developing the local culture and heritage industry, which is a key tourism attractor.

Finally, flows of tourists between home communities and destinations rely on networks of cars, highways, airplanes, and airports, as well as cruise ships or ferries (Urry 2008). As Cohen et al. (2011) note, "tourism is currently directly accountable for 4.4% of global CO2 emissions … with 40% of this figure conservatively attributed to tourist air travel" (Cohen et al. 2011, p. 1073). While tourism is intimately bound up with mobility networks, the associated environmental costs are often overlooked.

Global Climate Change as Social-Ecological Context

Our analytical focus is the oil-tourism interface in the North Atlantic. However, global climate change is the meta-context for both offshore oil extraction and nature-based tourism. Climate change is predicted to have significant impacts around the world, many of which are already in evidence. As noted by the IPCC (2018), climate change has significant implications for tourism economies in terms of where and when tourists travel, with some destinations (including many high latitude destinations) becoming more desirable, while others become less appealing.

Social and political responses to climate change also have ramifications if they limit air travel or other carbon-intensive forms of tourism mobility. Similarly, climate change has implications for offshore oil as global climate governance and national policymaking grapple with the need for low-carbon energy transitions. Ambitious climate change mitigation measures have the potential to amplify risks for governments and societies that depend on oil extraction for their economic wellbeing (IPCC 2014).

Climate change is talked about as a global environmental issue, but its effects are experienced at regional and local levels. Climate governance still largely happens in national political arenas. Across much of the North Atlantic, negative impacts are likely to include more severe autumn and winter storm seasons, increased rainfall and subsequent strain on infrastructure and flooding, and increasing sea levels and storm surges. For places like Iceland, climate change is causing glaciers to recede. This was dramatically marked in 2019 through a public "funeral" for the Okjökull glacier, which had shrunk to 6.6% of its original size and lost its glacier status in 2014 (Engel 2019). Conversely, while projected oceanic changes due to climate change suggest globally negative impacts for fisheries, high latitude regions may see increases in fish species richness and catch potential.

Our study regions share a North Atlantic geography. However, they are also characterized by Arctic or near-Arctic geographies. As the IPCC and others note, Arctic ecosystems and societies are disproportionately impacted by rapid and dramatic climate change effects that are already being felt, such as permafrost thawing and loss of sea ice (Callison 2014; Hall 2010). Cunsolo and Ellis (2018) note the increased unpredictability of sea ice has significant negative impacts on northern Indigenous community health because it disrupts the ability to go out on the land, resulting in a collective sense of "ecological grief" for severed connections to traditional practices and landscapes.

Changes to Arctic ecologies due to climate change are also seen to provide new opportunities. As the Arctic Ocean has longer ice-free seasons, it opens possibilities for oil or mineral exploration off the coasts of northern Norway, Iceland, or Greenland (Dodds 2010; Shadian 2014). For many Greenlandic Inuit, oil and mineral extraction is welcome if it is leveraged for greater political and economic independence from Denmark (Shadian 2014). In Iceland, by contrast, there is a pause in oil exploration activity due to oil price declines and volatility. This pause presents a decision point where Iceland can seek to join the oil countries at the risk of stranded assets, or else can choose to take an alternative path. Longer ice-free seasons have also spurred a growth in Arctic cruise ship travel, which trades on a desire to see Arctic landscapes and wildlife before they are transformed by climate change. Inuit communities in Labrador and Greenland, as well as Saami communities in Norway, are taking advantage of this increased interest by pursuing Indigenous-led tourism development.

The rapid and dramatic impacts of climate change on the Arctic underlie the emergence of a global Arctic as an object of scientific and political concern. That is, the Arctic is viewed less as the space of local communities, or the territorial jurisdiction of circumpolar countries. Rather, because Arctic climate change impacts global weather systems and oceanic cycles, it is increasingly subject to global scientific inquiry and political debate. The range of political and economic players that claim an interest in the Arctic is expanding to countries beyond the circumpolar region and arenas like the Arctic Council. China is becoming a significant presence in the Arctic with increasing activity related to natural resource exploration and extraction. Chinese-based oil company CNOOC-Nexen, for example, was involved in oil exploration off the northeast coast of Iceland. At the same time, China is a rapidly growing tourism market for Nordic and Arctic destination, which is a source of anxiety for some northern communities in terms of potential impacts of mass tourism. Parallel to this, international environmental organizations like WWF and Greenpeace are active in debates over Arctic oil exploration and extraction, targeting corporate and government actors across the north.

Climate change reshapes the oil-tourism interface in multiple ways, both in terms of new challenges, but also potentially new opportunities for both sectors. Rising sea levels and increasingly severe storm seasons can damage tourism infrastructure, transform coastlines, or damage historic sites that serve as tourism attractors. Changing fish and wildlife ranges could mean the movement or loss of wildlife species that serve as iconic tourism attractors in the region, such as whales, seals, puffins, or other seabirds. At the same time, if changing weather patterns result in shorter winters and longer summers this may mean extended tourism seasons. The expansion of Arctic cruise tourism may present opportunities for tourism companies and host communities in the region (James et al. 2020). Likewise, offshore oil sector infrastructure and oil hub cities, such as Esbjerg (Denmark), St. John's (Newfoundland and Labrador), Stavanger (Norway), and Aberdeen (Scotland) may be negatively impacted by increasingly severe storm seasons and sea level rise. Conversely, the decline of Arctic sea ice may open new frontiers for oil exploration, while longer ice-free seasons may reduce the risk of working in the harsh operating environment of the North Atlantic.

The oil-tourism interface also contributes to climate change. Globally, fossil fuels are significant sources of greenhouse gas emissions. John Urry (2013) offers the concept of the "carbon complex" to talk about the interconnected social actors and interests that maintain the oil sector. Most obviously, this includes the range of oil exploration and extraction companies, from large transnational operators like BP, Shell, or Exxon, to regional and local companies that provide the on-the-ground services at specific sites. But, as Urry (2013) notes, the carbon complex also involves a broad range of automobile, air travel, and shipping companies that are fossil fuel intensive and are closely connected to the oil sector. The carbon complex also involves national, regional, and local governments that benefit from oil royalties and the economic impacts of the sector. Finally, the carbon complex includes media and culture industries that benefit from the advertising revenues of oil and transportation companies, and which promote fossil fuel-intensive lifestyles of travel and consumerism. As societies that are intimately tied to fossil fuel extraction, Denmark, Newfoundland and Labrador, Norway, and Scotland are part of the carbon complex that contributes to climate change.

Attractive development is often lauded because it is a non-consumptive alternative to resource extraction, trading on the experience of travelling to new landscapes and communities. However, the tourism sector also contributes to climate change through its carbon intensity. Though less obvious than the systems of oil extraction and distribution at the core of the carbon complex, our dependence on fossil fuels involves several intersecting mobilities, including shipping miles for consumer goods, food miles embedded in agri-food distribution systems, and the personal travel required for work, to visit family and friends, and for leisure (Urry 2013). For some, the availability of accessible and affordable air travel contributes to a culture of "binge flying" that is problematic from a climate change perspective (Cohen et al. 2011). Car travel, cruise ships, boat tours, and other forms of motorized recreation also contribute to the carbon footprint of tourism.

Public and political debate around climate change also shapes the oil-tourism interface. The 2015 Paris Conference of the Parties (COP) meetings and Paris Agreement arguably represents a shift in the global climate governance regime, which is provoking new mitigation measures in many societies. Fossil fuel divestment movements, wherein institutional investors such as universities, churches, and municipalities remove their investments from fossil fuel companies, have also been successful in many places. These shifts lead to greater visibility for discussions about decarbonization and renewable energy transitions by scaling up solar, or onshore and offshore wind power. However, moves to decarbonization and adoption of renewables have not been without their own tensions. Some see the renewables transition as well-aligned with nature-based tourism as part of a green economy. However, there have also been conflicts over the landscape footprint of renewable energy, which can negatively impact wilderness values and Indigenous cultural practices, such as Saami reindeer herding, that form part of tourists' expectations and experiences. Responses to climate change vary across our research sites and will be explored in greater depth in Chapter 2.

Theoretical and Analytical Framework

This book is guided by a high-level theoretical framework that we call relational political ecology. Political ecology focuses on understanding the complex relationships among non-human environments, culture and discourse, and political economy. It includes a wide variety of more specific approaches as diverse as eco-Marxism, Actor–Network Theory, and post-structuralism (Bennett 2010; Escobar 1999; Latour 2004; Stoddart 2012), nearly all of which emphasize connections between the environment, politics, economy, and culture. Though we take inspiration from several approaches to political ecology, our *relational* political ecology is influenced by relational sociology (Crossley 2011; Emirbayer 1997). This relational approach leads us to emphasize the processes and networks involved in cultural and political deliberation, and formal and informal politics such as governance processes and social movement interventions. This approach adopts a model of social power that emphasizes how power works through inclusion or exclusion from arenas of deliberation and decision-making, as well as centrality, reputation, and influence within these arenas (Castells 2009; Crossley 2011; Di Gregorio et al. 2019; Fischer and Sciarini 2015). While state actors are particularly important "network hubs," non-state players, including publics, social movements, and corporations, exercise power through multiple "media of exchange," including visibility through mass media and social media, the "political technologies" of protest and voting, or by exerting economic influence (Crossley 2011). Individuals and social groups also exercise power and influence through networks of collaboration and alignment with others on shared eco-political projects, or by being disruptive and creating friction through networks of conflict and contention.

Figure 1.1 summarizes this way of thinking about political economy and culture as embedded in broader ecological contexts. Governance processes and social movement activities unfold over time in multiplex networks, which are composed of multiple overlapping networks from different domains (cultural, political economic, and ecological). In addition to emphasizing how culture and political economy are embedded in ecologies, we show how the main divisions of cases in terms of their political economies and dominant "orders of worth"—introduced below and discussed in detail in Chapter 3—are roughly homologous. For example, Norway and Newfoundland and Labrador are relatively high-oil orientation and relatively low-tourism orientation in the political economy space (at the top of the figure). They also occupy a similar position in the cultural space, which in this case is defined in terms of a strong orientation towards "industrial worth" and a relatively low

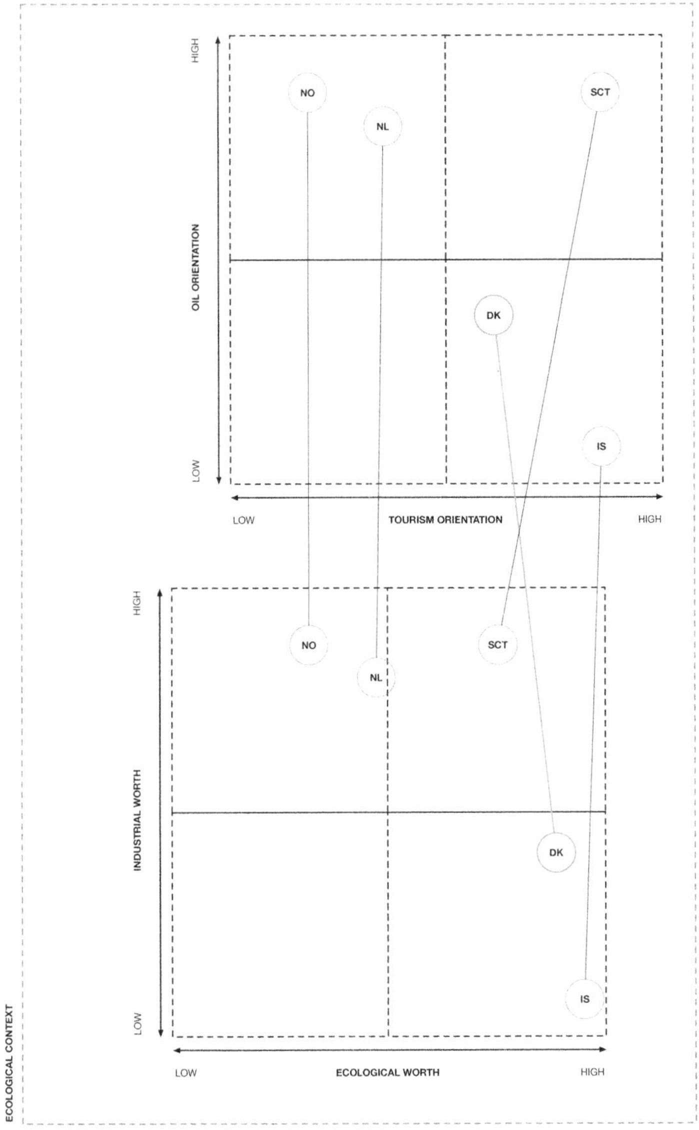

Fig. 1.1 Relational political ecology

orientation to "ecological worth," each of which is defined and discussed below. The key idea is that these are roughly homologous spaces that are embedded in larger ecologies. Together they make up the structural contexts where governance (Chapter 4) and social movement strategic interaction (Chapter 5) take place.

The ecological contexts we have in mind—in addition to global climate change, discussed previously—are made up of oceans, beaches, headlands, maritime barrens, forests, birds, and wildlife, as well as national parks, and protected areas. They include flows of oil that fuel tourism mobility, as well as the carbon emissions produced by the offshore oil industry, which draws oil from beneath the seabed as part of the global oil industry. Together, they represent the outermost context of Fig. 1.1.

One of our key ideas is that we should focus on the direct and indirect *connections* between extractive and attractive development. We can do so by thinking about the multiplex political and cultural processes and networks that are embedded in specific political economies and ecologies. This is our relational approach to political ecology. Ultimately, it is about bringing a relatively neglected set of issues into focus and thinking about them.

Our thinking about ecological relationships is also informed by an emerging body of research variously labelled sociology of the oceans or marine sociology (Hannigan 2016; Longo and Clark 2016; Widener 2018). Oceans play multiple vital historic and contemporary roles in social-ecological relationships, yet receive limited attention in sociology, including in the subdiscipline of environmental sociology (Urry 2014). There are three important facets of these relationships. First, oceans serve as pathways for carbon-intensive shipping of consumer goods and food. Oceans are also pathways for tourist and leisure travel, including long-distance cruise ship travel that goes far offshore, but also short-distance and nearshore travel like boat tours or sea kayaking (Urry 2014). Second, oceans are treated as a resource, which includes offshore oil, but also fisheries and aquaculture, as well as emerging projects like deep-sea mining (Hannigan 2016). Third, oceans are sites of ecological problems and objects of political contestation and social movement mobilization. Atmospheric climate change intersects with ocean ecologies to create more intense hurricanes and typhoons, rising sea levels, Arctic and Antarctic ice melt, and shifting ranges of marine species. As climate change makes polar regions increasingly accessible as a pathway for shipping and cruise ship travel, there are also emerging conflicts over Heavy Fuel Oil (HFO) use in ecologically sensitive Arctic waters. Other key issues include ocean plastics and global fisheries declines, which have substantial localized impacts. Collective action around coastal pollution has been based on the recreational use of these environments, such as surfing-based environmental movements like Surfers Against Sewage (Wheaton 2008). International environmental organizations, such as Greenpeace and the World Wildlife Fund, increasingly oppose oceanic resource extraction (Hannigan 2016; Widener 2018). This includes opposition to offshore oil extraction, particularly as it extends into Arctic environments (Stoddart et al. 2019).

As Fikret Berkes (2015) notes, building governance processes that contribute to the social-ecological wellbeing of coastal societies and environments is complicated because of the multi-scalar qualities of oceans. Coastal environments have different types of zones, including the landward zone, the shoreline, the nearshore, and the offshore. These zones are subjects to different, sometimes overlapping, political jurisdictions, and governance regimes. This creates significant complexity in terms of achieving the social or political changes needed to address the problems of oceans and transition to more sustainable social-ecological relationships.

In the context of the oil-tourism interface, much coastal tourism happens on the landward side, or along the shoreline, as this is where cities and towns are located (along with hotels and restaurants) and where popular activities like coastal hiking take place. The oil sector inhabits these spaces primarily through office buildings, as spaces for oil transportation and refining, or as harbours for offshore service vessels. The nearshore is used for nature-based tourism practices like boat tours and sea kayaking but can also be the site of offshore oil infrastructure. The nearshore can be part of the pathways of cruise ship travellers as they enter or leave port, or for oil industry service vessels. By contrast, the offshore zone is where offshore oil extraction takes place and is out of view of tourism activity in the landward, shoreline, or nearshore zones. Instead, tourist experiences of the offshore are generally limited to cruise ship travel. The complexities of coastal ecologies, social uses, political jurisdictions, and governance regimes are important considerations for our analysis of the governance, social movements, and cultural dimensions of the oil-tourism interface.

The substantive chapters of the book contribute to this high-level relational political ecology perspective by using a combination of specific conceptual frameworks to help us answer the empirical questions introduced earlier. First, the players and arenas framework emerged as a critique of political process models and opportunity structures in the social movements literature (Duyvendak and Jasper 2015; Jasper and Duyvendak 2015). This framework emphasizes strategic interaction between opposing players within a constrained and evolving environment. The framework provides guiding questions about what players want (e.g. individuals as well as compound players such as movements, counter-movements, organizations, governments), what resources they have available, what constraints they operate within, and how shared or stable their goals are. These players interact strategically within arenas, which vary in their degree of institutionalization and regulation (e.g. public opinion versus the courts).

Second, we draw on theory and research on evaluative repertoires and institutionalized "orders of worth" from comparative cultural sociology (Boltanski and Thévenot 2006). The key concepts here—evaluative repertoires and orders of worth—were initially developed in conversation with Pierre Bourdieu's work on distinction, social inequality, and cultural cognition (i.e. the habitus), and as a critique of essentialist approaches to cultural comparison. As we show in Chapter 3, the shift to comparing how tourism and oil are *evaluated* reveals similarities and differences in what is most valued and legitimated across societies. It reveals, for example, when and where ecological ways of thinking and valuing are more visible than market or industrial thinking.

While these concepts of evaluative repertoires and orders of worth have developed in different scholarly traditions, these concepts are deeply compatible with the players and arenas framework. Both frameworks begin from relaxing assumptions that were central to Bourdieu's theoretical framework. While the players and arenas framework is primarily a critique of political process theory, it also developed in part by relaxing assumptions about the constraints on action imposed by the structural features of social and political fields. Similarly, the literature on evaluative repertoires and orders of worth emerged by relaxing assumptions about the institutional logics of *distinction* embedded in class inequalities and cultural fields, and of the overly deterministic nature of habitus as a model for cultural cognition. In part because of these common origins, the pairing of the players and arenas framework with evaluative repertoires and orders of worth enables us to compare not only how our case study regions assess the relative merits of oil and tourism, but also how those distinctions are strategically disputed and governed.

In this book, and because of our interest in *relational* political ecology, we attend to the connections and lack of connections between ideas that are invoked in cultural networks (DiMaggio 2011; Pachucki and Breiger 2010; Rule and Bearman 2015). When we talk about culture—for example in news stories—do we connect extractive and attractive development into common conversations, or do we keep them separate? Do our collective cultural repertoires within and across our case study regions differ in the cultural bridges and "holes" between these two modes of development? To answer these questions, we combine the general idea of cultural repertoires with cultural approaches to network analysis, specifically the idea of cultural holes (Pachucki and Breiger 2010). As we show in Chapter 3, cultural holes are the general pattern that holds across all five cases. Public deliberation of the merits of oil and tourism rarely intersect in political or media arenas.

Our general orientation is one of relational political ecology, while our more specific theories—players and arenas and evaluative repertoires and orders of worth—concern the cultural and political acts that connect or separate these two extractive and attractive development models. The larger goal, to which we turn now, is understanding how societies can navigate these modes of development to increase social-ecological wellbeing.

The Structure of the Book

To answer our research questions, we conducted mixed-methods research in a range of sites across the North Atlantic region, including in Denmark, Iceland, Newfoundland and Labrador, Norway, and Scotland. A discussion of why we take the North Atlantic as our object of inquiry is presented in Chapter 2. Our data include interviews with representatives from the energy sector, tourism sector, various levels of government, and environmental organizations. Our data also includes fieldwork and observation at sites across the region, including at events oriented around offshore oil or tourism development. We also draw on a range of texts and documents in our analysis, including policy documents, mass media articles, organizational websites, and web 2.0 content. The Epilogue on Methodology provides a methodological overview of our data collection and analysis.

The Epilogue on Methodology also sets out the methodological standpoint of the project, which builds from the concept of methodological omnivorousness. Omnivores are animals (including humans) that subsist on plant-based and animal-based food. Besides humans, bears offer a good metaphorical image of omnivorousness. Bears adopt a diet of fish, berries, insects, or plants depending on season and availability. As such, the concept of methodological omnivorousness helps us think through the potential of mixed-methods approaches to address complex research problems. While the three of us approach this project with different methodological backgrounds, this methodological standpoint serves as an intellectual co-working space.

Chapter 3 builds on the political economy framework presented in Chapter 2 with a comparison of how oil and tourism development are evaluated across our five cases. This focus on evaluative repertoires (Lamont 2012; Lamont and Thévenot 2000; Tilly 2006) and concomitant boundary processes (Lamont and Molnár 2002) reveal similarities and differences in what is valued (e.g. conservation, jobs, energy security). More specifically, it sheds light on the cultural *hierarchies* of value in each of our cases. Following Boltanski and Thévenot (1999, 2006), we refer to these institutionalized cultural hierarchies as "orders of worth." Across cases, we see that ecological orders of worth—in which value, or "worthiness," emphasizes ecology, nature conservation, and responses to climate change—are more strongly associated with tourism, while industrial orders of worth—emphasizing scientific and technical innovation and efficiency—are more strongly associated with oil development. Market orders of worth are also common across all our cases. Despite these cross-case similarities, we find differences in the relative weight afforded to each order of worth. These variations are roughly homologous with the political economy orientations we describe in Chapter 2.

Chapter 4 examines environmental governance related to the oil-tourism interface. We draw on research on the shift from government as the key source of environmental policymaking to a broader understanding of governance. While governments remain powerful, broader networks of social actors, including corporations, social movements, scientists, and other experts increasingly have a role in shaping environmental politics and policy. This dimension of environmental governance may be thought of as "horizontal," in that it stretches out to include more diverse players (Francesch-Huidobro 2012). Environmental governance also increasingly involves action across local, sub-national, national, or international political arenas (Lövbrand et al. 2017; Vasi 2007). This dimension of environmental governance is "vertical," in that it includes organizations and other collective players from small local political arenas to massive global ones (Francesch-Huidobro 2012). Environmental governance research draws attention to both the horizontal and vertical dimensions in terms of how participatory or restrictive governance processes are, who is permitted or who has sufficient resources to participate, who sets the terms of reference and how these factors shape outcomes (Adkin 2009). Unlike much research in the environmental governance literature, our relational political ecology approach foregrounds the connections (or lack of connections) across sectors and environmental issues. Governance related to different business sectors, industrial activity, or forms of land use is often pursued as *separate processes*, siloed off from each other. However, as environmental scientists increasingly highlight, different forms of development intersect to have cumulative landscape-level impacts. This has led to interest in models of landscape-level planning and governance. With our own interests in oil extraction and tourism development, we are most interested in if, when and how the connective dimension of environmental governance brings these modes of development together in shared political or planning arenas. The connective dimension of environmental governance may also involve links between oil extraction or tourism development and other forms of coastal development, including fisheries or renewable energy development.

Chapter 5 casts light on when and how environmental movements intervene in the oil-tourism interface. Social movements and non-governmental organizations are particularly important players in environmental governance. Environmental movements intervene in public and political arenas on behalf of citizens and to speak on behalf of nature. As Castells (2004) notes, since the 1960s protest cycle, environmental movements have shifted political discourse and public opinion about human–environment relationships. Sustainability is now integrated throughout political and corporate discourse to the point where it becomes difficult to assess whether the language of sustainability is anything more than an empty signifier. Environmental movements, like other social movements, rely on a multifaceted "repertoire of communication" involving multiple media channels to reach different audiences that include journalists, bystander publics, politicians, and sympathetic bystanders who may join the movement (Mattoni 2012). Social movement scholars have noted a relationship of asymmetrical dependence between social movements and the mass media, wherein movements depend on the media for coverage and visibility more than the media depend on movements for content (Carroll and Ratner 1999). This leads many environmental groups, most notably Greenpeace, to embrace dramatic and spectacular actions that create media-friendly imagery. The pursuit of media coverage through dramatic social movement action involves a trade-off, where quantity of media coverage often does not create quality coverage in terms of conveying environmental movement claims and grievances (Hutchins and Lester 2006). Furthermore, media coverage reinforces perceptions of moderate environmental groups and tactics as politically legitimate and socially acceptable, while radical forms of eco-activism are delegitimated (Newlands 2018). As many have noted, the advent of web-based communication technologies makes it easier for movement players to circumvent journalists and editors in reaching audiences through their own webpages, Facebook, YouTube, Instagram, or Twitter feeds (Katz-Kimchi and Manosevitch 2015; Lester and Hutchins 2009; Newlands 2018). While this levels the traditionally asymmetrical relationship to a certain degree, mainstream media coverage by flagship or legacy news outlets remains important for reaching bystanders who are not already interested in or supportive of movement activities (Castells 2009; Hutchins and Lester 2015). Similarly, mainstream media coverage can be effective in leveraging the political efficacy of social movements as they engage in the political sphere.

After answering our questions about cultural variation and evaluative repertoires, environmental governance, and social movement strategic interaction, we return to our overarching question about how societies in the North Atlantic may better navigate the oil-tourism interface. In Chapter 6, we discuss the broader implications of our research for environmental social scientists concerned with how societies navigate relationships between extractive and attractive development. We also offer lessons learned from this research that may inform policy and practice by policymakers, community leaders, social movements and civil society organizations, and stakeholders in the energy sector or tourism sector.

Our conclusions highlight that offshore oil and tourism are generally treated as separate development paths, though both are often interpreted as positive for host communities in terms of economic development and social benefits. However, our analysis of contact points across sectors shows that they do not simply coexist but are intertwined in a variety of ways, which can be direct or indirect, conflictual, or collaborative. The coexistence of these sectors is often taken for granted in host communities, until new projects for oil exploration or extraction infringe on established and valued tourism economies. This can provoke conflict between different ways of valuing nature and integrating it into local economies. As our analysis shows, environmental movement and civil society players intervene in the oil-tourism interface in diverse ways that involve both conflict and collaboration.

A key conclusion of this project is to encourage more connective ways of thinking about, planning for, and managing the multiple modes of development that share social and ecological space in coastal communities. Connective forms of environmental governance should be inclusive of non-governmental and civil society players because they play a vital role in envisioning new social-ecological futures for coastal communities. This is consistent with other calls in the environmental sciences and social sciences for greater attention to landscape-level (or seascape-level) planning and management, or for greater attention to cumulative environmental impacts across multiple forms of development. Instead of waiting until periods conflict to recognize the connections between offshore oil and tourism, we believe that creating more connective spaces for linking viewpoints and interests from different sectors will contribute to building social-ecological wellbeing for coastal communities.

References

Adkin, L. E. (Ed.). (2009). *Environmental conflict and democracy in Canada*. Vancouver, BC: UBC Press.

Bennett, J. (2010). *Vibrant matter: A political ecology of things*. Durham, NC: Duke University Press.

Berkes, F. (2015). *Coasts for people: Interdisciplinary approaches to coastal and marine resource management*. New York: Routledge.

Berkes, F., & Ross, H. (2013). Community resilience: Toward an integrated approach. *Society & Natural Resources, 26*(1), 5–20.

Boltanski, L., & Thévenot, L. (1999). The sociology of critical capacity. *European Journal of Social Theory, 2*(3), 359–377.

Boltanski, L., & Thévenot, L. (2006). *On justification: Economies of worth* (C. Porter, Trans.). Princeton, NJ: Princeton University Press.

Callison, C. (2014). *How climate change comes to matter: The communal life of facts*. Durham, NC: Duke University Press.

Carroll, W. K., & Ratner, R. S. (1999). Media strategies and political projects: A comparative study of social movements. *Canadian Journal of Sociology, 24*(1), 1–34.

Castells, M. (2004). *The power of identity* (2nd ed.). Malden, MA: Blackwell Publishing.

Castells, M. (2009). *Communication power*. Oxford: Oxford University Press.

Cohen, S. A., Higham, J. E. S., & Cavaliere, C. T. (2011). Binge flying: Behavioural addiction and climate change. *Annals of Tourism Research, 38*(3), 1070–1089.

Crossley, N. (2011). *Towards relational sociology*. London: Routledge.

Cunningham, P. A., Huijbens, E. H., & Wearing, S. L. (2012). From whaling to whale watching: Examining sustainability and cultural rhetoric. *Journal of Sustainable Tourism, 20*(1), 143–161.

Cunsolo, A., & Ellis, N. R. (2018). Ecological grief as a mental health response to climate change-related loss. *Nature Climate Change, 8*(4), 275–281.

Davidson, D. J. (2010). The applicability of the concept of resilience to social systems: Some sources of optimism and nagging doubts. *Society and Natural Resources, 23*(12), 1135–1149.

Di Gregorio, M., Fatorelli, L., Paavola, J., Locatelli, B., Pramova, E., Nurrochmat, D. R., et al. (2019). Multi-level governance and power in climate change policy networks. *Global Environmental Change, 54*, 64–77.

DiMaggio, P. (2011). Cultural networks. In J. Scott & P. J. Carrington (Eds.), *The Sage handbook of social network analysis* (pp. 286–300). London: Sage.

Dodds, K. (2010). A polar Mediterranean? accessibility, resources and sovereignty in the Arctic Ocean. *Global Policy, 1*(3), 303–311.

Duyvendak, J. W., & Jasper, J. M. (2015). *Breaking down the state: Protestors engaged*. Amsterdam: Amsterdam University Press.

Emirbayer, M. (1997). Manifesto for a relational sociology. *American Journal of Sociology, 103*(2), 281–317.

Engel, C. (2019). Scientists unveil memorial to Iceland's 'first' dead glacier lost to climate change. *Time*. https://time.com/5631599/iceland-glacier-climate-change/. Accessed 22 July 2019.

Escobar, A. (1999). After nature: Steps to an antiessentialist political ecology. *Current Anthropology, 40*(1), 1–30.

Fischer, M., & Sciarini, P. (2015). Unpacking reputational power: Intended and unintended determinants of the assessment of actors' power. *Social Networks, 42*, 60–71.

Francesch-Huidobro, M. (2012). Institutional deficit and lack of legitimacy: The challenges of climate change governance in Hong Kong. *Environmental Politics, 21*(5), 791–810.

George, E. W., Mair, H., & Reid, D. G. (2009). *Rural tourism development: Localism and cultural change*. Bristol, UK: Channel View Publications.

Gould, K. A. (1999). Tactical tourism: A comparative analysis of rainforest development in Ecuador and Belize. *Organization & Environment, 12*(3), 245–262.

Gould, K. A. (2017). Ecotourism under pressure: The political economy of oil extraction and cruise ship tourism threats to sustainable development in Belize. *Environmental Sociology, 3*(3), 237–247.
Gudynas, E. (2011). Buen Vivir: Today's tomorrow. *Development, 54*(4), 441–447.
Hall, C. M. (2010). Tourism and environmental change in polar regions: Impacts, climate change and biological invasion. In C. M. Hall & J. Saarinen (Eds.), *Tourism and change in polar regions: Climate, environments and experiences* (pp. 42–70). London: Routledge.
Hannigan, J. (2016). *The geopolitics of deep oceans.* Cambridge, UK: Polity Press.
Hutchins, B., & Lester, L. (2006). Environmental protest and tap-dancing with the media in the information age. *Media, Culture and Society, 28*(3), 433–451.
Hutchins, B., & Lester, L. (2015). Theorizing the enactment of mediatized environmental conflict. *International Communication Gazette, 77*(4), 337–358.
IPCC. (2014). *Climate change 2014: Synthesis report: Contribution of working groups I, II and III to the fifth assessment report of the Intergovernmental Panel on Climate Change.* Geneva, Switzerland: IPCC.
IPCC. (2018). *Global warming of 1.5°C: An IPCC special report on the impacts of global warming of 1.5°C above pre-industrial levels and related global greenhouse gas emission pathways, in the context of strengthening the global response to the threat of climate change, sustainable development, and efforts to eradicate poverty.* Geneva, Switzerland: IPCC.
James, L., Olsen, L. S., & Karlsdóttir, A. (2020). Sustainability and cruise tourism in the Arctic: Stakeholder perspectives from Ísafjörður, Iceland and Qaqortoq. *Greenland. Journal of Sustainable Tourism, 28*(9), 1425–1441.
Jasper, J. M., & Duyvendak, J. W. (2015). *Players and arenas: The interactive dynamics of protest.* Amsterdam: Amsterdam University Press.
Jennings, A. (2015). The old rock's rockin': A cultural anatomy of Shetland. In K. Vodden, R. Gibson, & G. Baldacchino (Eds.), *Place peripheral: Place-based development in rural, island, and remote regions* (pp. 156–179). St. John's, NL: Memorial University, Institute of Social and Economic Research.
Katz-Kimchi, M., & Manosevitch, I. (2015). Mobilizing Facebook users against Facebook's energy policy: The case of Greenpeace unfriend coal campaign. *Environmental Communication, 9*(2), 248–267.

Lamont, M. (2012). Toward a comparative sociology of valuation and evaluation. *Annual Review of Sociology, 38,* 201–221.
Lamont, M., & Molnár, V. (2002). The study of boundaries in the social sciences. *Annual Review of Sociology, 28*(1), 167–195.
Lamont, M., & Thévenot, L. (2000). *Rethinking comparative cultural sociology: Repertoires of evaluation in France and the United States.* Cambridge, UK: Cambridge University Press.
Latour, B. (2004). *Politics of nature: How to bring the sciences into democracy* (C. Porter, Trans.). Cambridge, MA: Harvard University Press.
Laudati, A. (2010). Ecotourism: The modern predator? Implications of gorilla tourism on local livelihoods in Bwindi Impenetrable National Park, Uganda. *Environment and Planning D: Society and Space, 28*(4), 726–743.
Lester, L., & Hutchins, B. (2009). Power games: Environmental protest, news media and the internet. *Media, Culture and Society, 31*(4), 579–595.
Longo, S. B., & Clark, B. (2016). An Ocean of troubles: Advancing marine sociology. *Social Problems, 63*(4), 463–479.
Lövbrand, E., Hjerpe, M., & Linnér, B. (2017). Making climate governance global: How UN climate summitry comes to matter in a complex climate regime. *Environmental Politics, 26*(4), 580–599.
Luke, T. W. (2002). On the political economy of Clayoquot Sound: The uneasy transition from extractive to attractive models of development. In W. Magnusson & K. Shaw (Eds.), *A political space: Reading the global through Clayoquot Sound* (pp. 91–112). Montreal and Kingston: McGill-Queen's University Press.
Luzecka, P. (2016). "Take a gap year!" A social practice perspective on air travel and potential transitions towards sustainable tourism mobility. *Journal of Sustainable Tourism, 24*(3), 446–462.
Mattoni, A. (2012). *Media practices and protest politics: How precarious workers mobilise.* Surrey, UK: Ashgate.
Newlands, M. (2018). *Environmental activism and the media: The politics of protest.* New York: Peter Lang.
Pachucki, M. A., & Breiger, R. L. (2010). Cultural holes: Beyond relationality in social networks and culture. *Annual Review of Sociology, 36,* 205–224.
Reed, M. G., & Gill, A. M. (1997). Tourism, recreational, and amenity values in land allocation: An analysis of institutional arrangements in the postproductivist era. *Environment and Planning A, 29*(11), 2019–2040.

Reichwein, P. (2014). *Climber's paradise: Making Canada's mountain parks, 1906–1974*. Edmonton, AB: University of Alberta Press.

Rule, A., & Bearman, P. (2015). Networks and culture. In L. Hanquinet & M. Savage (Eds.), *Routledge International handbook of the sociology of art and culture* (pp. 161–173). London: Routledge.

Shadian, J. M. (2014). *The politics of Arctic sovereignty: Oil, ice, and Inuit governance*. London: Routledge.

Shaw, K., & Magnusson, W. (Eds.). (2002). *A political space: Reading the global through Clayoquot Sound*. Montreal and Kingston: McGill-Queen's University Press.

Sheller, M., & Urry, J. (Eds.). (2004). *Tourism mobilities: Places to play, places in play*. London: Routledge.

Snyman, S., & Bricker, K. S. (2019). Living on the edge: Benefit-sharing from protected area tourism. *Journal of Sustainable Tourism, 27*(6), 705–709.

Stoddart, M. C. J. (2012). *Making meaning out of mountains: The political ecology of skiing*. Vancouver, BC: UBC Press.

Stoddart, M. C., Chewinski, M., Burt, B. Q., & Stewart, M. (2019). Political consumerism in the oil and mining extractive industries: Possibilities for sustainability and social justice. In M. Bostrom, M. Micheletti, & P. Oosterveer (Eds.), *The Oxford handbook of political consumerism* (pp. 367–388). New York: Oxford University Press.

Szerszynski, B., & Urry, J. (2006). Visuality, mobility and the cosmopolitan: Inhabiting the world from afar. *The British Journal of Sociology, 57,* 113–131.

Tilly, C. (2006). *Why? What happens when people give reasons ... and why*. Princeton, NJ: Princeton University Press.

Thrift, N. (2000). Still life in nearly present time: The object of nature. *Body & Society, 6*(3–4), 34–57.

Uldam, J. (2018). Social media visibility: Challenges to activism. *Media, Culture and Society, 40*(1), 41–58.

Urry, J. (2008). Climate change, travel and complex futures. *British Journal of Sociology, 59,* 261–279.

Urry, J. (2013). *Societies beyond oil: Oil dregs and social futures*. London: Zed Books.

Urry, J. (2014). *Offshoring*. Cambridge, UK: Polity Press.

Urry, J., & Larsen, J. (2011). *The tourist gaze 3.0*. Los Angeles: Sage.

Vasi, I. B. (2007). Thinking globally, planning nationally and acting locally: Nested organizational fields and the adoption of environmental practices. *Social Forces, 86*(1), 113–136.

Waitt, G., & Cook, L. (2007). Leaving nothing but ripples on the water: Performing ecotourism natures. *Social and Cultural Geography, 8*(4), 535–550.

Wheaton, B. (2008). From the pavement to the beach: Politics and identity in 'Surfers Against Sewage'. In M. Atkinson & K. Young (Eds.), *Tribal play: Subcultural journeys through sport* (pp. 113–134). Bingley, UK: JAI Press.

Whitmee, S., Haines, A., Beyrer, C., Boltz, F., Capon, A. G., Dias, B. F. d. S., et al. (2015). Safeguarding human health in the Anthropocene epoch: Report of The Rockefeller Foundation–*Lancet* Commission on planetary health. *Lancet, 386*, 1973–2028.

Widener, P. (2009). Oil tourism: Disasters and destinations in Ecuador and the Philippines. *Sociological Inquiry, 79*(3), 266–288.

Widener, P. (2011). *Oil injustice: Resisting and conceding a pipeline in Ecuador*. Lanham, MD: Rowman & Littlefield Publishers.

Widener, P. (2018). Coastal people dispute offshore oil exploration: Toward a study of embedded seascapes, submersible knowledge, sacrifice, and marine justice. *Environmental Sociology, 4*(4), 405–418.

2
The North Atlantic as Object of Inquiry

We orient around the North Atlantic as our object of inquiry, exploring cultural, governance, and social movements dimensions of the oil-tourism interface across Denmark, Iceland, Newfoundland and Labrador (Canada), Norway, and Scotland (United Kingdom). This regional focus on the North Atlantic is a response to calls to move beyond the methodological nationalism that treats nation-states as the natural, de facto units of analysis in comparative research, which has dominated social sciences for decades (i.e. Beck and Levy 2013; Thorpe 2012). At the same time, a focus on cases that are linked through the shared maritime geography of the northern Atlantic Ocean extends recent sociological research on oceans as sites of conflict over resource development (Hannigan 2017; Longo and Clark 2016; Widener 2018).

Our case study regions can be conceptualized in several ways. Scandinavia includes Norway and Denmark, as well as Sweden, while the broader framework of the Nordic countries (or *Norden*) also includes Iceland and Finland. Newfoundland and Labrador is a province of Canada, while Scotland is a more autonomous region within the United Kingdom. Denmark is part of the European Union. Scotland

M. C. J. Stoddart et al., *Industrial Development and Eco-Tourisms*,
https://doi.org/10.1007/978-3-030-55944-1_2

is embroiled in the Brexit process with the UK leaving the European Union, which has renewed interest in Scottish independence. The Kingdom of Denmark includes Denmark, but also Greenland and the Faroe Islands, which receive limited attention here. Furthermore, except for Scotland, the case study regions have Arctic or near-Arctic geographies and are represented on the Arctic Council. Despite these different—and sometimes overlapping—geographies and political affiliations, it is useful to group them within a North Atlantic regional framework for several reasons.

First, previous comparative research examines various configurations of these case study regions. The edited volume, *Northern Neighbours: Scotland and Norway since 1800*, looks at how issues like access to nature and outdoor recreation, social welfare, and oil development have been shaped by long-term cultural, political, and economic similarities and differences between Scotland and Norway (Bryden et al. 2015a). An overarching theme of the book is that the Norwegian model of governance and natural resource development, characterized by a "more egalitarian, healthy and participatory society," offers lessons for Scotland (Bryden et al. 2015b, p. 19). The edited volume, *Nordic Tourism: Issues and Cases* examines a range of northern tourism issues, including environmental sustainability and rural economic development, through an analysis of Norway, Denmark, and Iceland (Hall et al. 2009). Wejs et al. (2014) examine the challenges of building legitimacy for climate change adaptation plans through a comparison of Denmark and Norway. Søholt et al. (2018) compare Denmark and Norway in their study of the social dynamics of immigration in rural Nordic communities. Felt (2009) examines structures of municipal governance in the small cities of Corner Brook, on the west coast of Newfoundland, and Akureyri, on the north coast of Iceland, and argues that there are long-term cultural and political characteristics that have resulted in Iceland adopting more pluralistic forms of governance than are found in Newfoundland and Labrador.

Second, our analysis complements other moves to build communities of research, policy, and practice around the North Atlantic region. An example of this is the North Atlantic Forum, a biannual conference that has been held in Atlantic Canada, Iceland, and Norway, which has led to a series of edited volumes that consolidate research on the North Atlantic (Baldacchino et al. 2009, 2015; Brinklow and Gibson 2017). These volumes touch on issues including oil development, tourism, and governance, with a focus on how rural communities across the North Atlantic navigate the challenges of globalization and rapid social-ecological change.

Third, these study regions are linked by broad similarities of geography, relatively low population density, and democratic political systems. There has also been an historical dependence on fisheries, with more recent turns to offshore oil and nature-based tourism as alternative economic development pathways. Yet, there are significant differences in how well-established and economically significant the tourism and oil industries are in the different sites. For example, while oil exploration and extraction took off in the 1970s among the North Sea cases (Denmark, Norway, and Scotland), oil did not become central to the political economy of Newfoundland and Labrador until the early 2000s, while Iceland was at the early stages of exploration in the Dragon oil field when the 2014 global oil price collapse resulted in cooling interest in Arctic oil exploration. By contrast, Scotland has a long history as a nature-based tourism destination, while Denmark and Newfoundland and Labrador have more recently turned to tourism development as a means of economic diversification. Iceland, by contrast, experienced a recent and rapid tourism boom.

Our cases represent a range of political and economic orientations. According to Lijphart's (2012) typology of forms of democracy, Canada and the United Kingdom are at the majoritarian and pluralist end of the political spectrum where the winning political party exercises a great deal of policymaking power. Denmark, Iceland, and Norway, by contrast, are at the consensus-oriented and corporatist end of the spectrum, with political cultures characterized by cross-party alliances and higher levels of consultation with non-state actors. Similarly, our cases contain a mix of liberal market economies (Canada, the United Kingdom) that rely more heavily on markets to structure "financial and industrial relations systems" and coordinated market economies (Denmark, Norway) that embrace "higher levels of non-market coordination" (Hall and Soskice 2001, p. 19). The combination of similarities and differences allows for comparative work that helps us better understand when and how nature-oriented tourism and offshore oil come into contact.

In the remainder of this chapter, we set the stage for our analysis by providing an overview of the political and economic context of oil and tourism development in our five case study regions. We then provide an overview of how our case study regions are responding to the challenges of climate change, as this an important part of the meta-context of the oil-tourism interface in the North Atlantic.

Oil Development in the North Atlantic

North Atlantic oil exploration started in the 1960s in the North Sea around the Scottish, Norwegian, and Danish continental shelves. The 1970s OPEC oil crises was a tipping point in the political economy of offshore oil extraction for the North Atlantic. The OPEC crisis was triggered by the coalition of Middle Eastern oil extracting countries and results in price spikes and oil shortages. A key outcome of the crisis was an emphasis on "energy independence" from the OPEC countries and "energy security" in North America and Europe (Freudenburg and Gramling 2011; Kristoffersen 2014; Urry 2013). As a result, there was a turn to what John Urry (2013) terms "tough oil." In contrast to the "easy oil" that is extracted from land-based derricks in places like the Middle East or Texas, tough oil is in more remote locations, has a lower ratio of energy returned on energy invested (EROEI), and carries higher environmental risks and impacts. Ramping up oil exploration and extraction across the North Atlantic was part of this shift towards tough oil in response to increasing global oil prices and the search for energy security through the 1970s.

Table 2.1 provides indicators that summarize how oil extraction fits into the political economies of our study regions. Looking at oil sector employment as a percentage of the population, we see that Newfoundland and Labrador (0.99%) and Norway (0.95%) have the highest level of employment in the sector, followed by Scotland (0.66%). By contrast, oil sector employment as a percentage of the population is much lower in Denmark (0.09%). Data is unavailable for Iceland but given the

Table 2.1 Oil orientations of case study regions

	Oil sector employment (2018, unless otherwise noted)	Population (2018)	Oil sector employment as % of population	Proven oil reserves, thousand million barrels, (2018)[a]	Oil production, in million tonnes (2018, unless otherwise noted)[b]	Growth rate per annum in oil production, in million tonnes (2017–2018)[c]	Oil products exports (value in billion US dollar) (2018, unless otherwise noted)
Denmark	5585 (2017)[d]	5,781,190[e]	0.09	0.4	5.7	-15.9%	4.19[f]
Iceland	n/a	357,050[g]	n/a	n/a	n/a	n/a	n/a
Norway	50,600[h]	5,295,619[i]	0.95	8.6	83.1	-6.2%	51.18[j]
Newfoundland and Labrador	5,200[k]	525,075[l]	0.99	3[m]	11.2[n]	4.3%[o]	5.92[p]
Scotland	36,100 (2017)[q]	5,438,100[r]	0.66	2.5 (UK)	48.7 (2017)[s]	0.5%[t]	20.51(2017)[u]

[a]BP Statistical Review of World Energy, 2019
[b]BP Statistical Review of World Energy, 2019
[c]BP Statistical Review of World Energy, 2019
[d]ERHV1 table, Statistics Denmark, 2019 (full-time employment in extraction of crude petroleum, extraction of natural gas, support activities for petroleum and natural gas mining, manufacture of refined petroleum products, and wholesale of solid, liquid and gaseous fuels and related products)
[e]Statistics Denmark, 2019
[f]UHV7 table, Statistics Denmark, 2019, conversion rate: 1DKK = 0.15358 USD
[g]Statistics Iceland, 2018
[h]Statistics Norway, 2019

[i]Statistics Norway, 2019
[j]Norwegian Petroleum, 2019, conversion rate: 1NOK = 0.1158 USD
[k]Economy 2019, the Department of Finance, Newfoundland & Labrador government, 2019
[l]Statistics Canada, 2019
[m]The Department of Natural Resources, Newfoundland & Labrador government, 2019
[n]Newfoundland & Labrador Statistics Agency, the Department of Finance, Newfoundland & Labrador government, 2019, unit conversion: 1 tonne of crude oil = 7.5 barrels of oil equivalent
[o]Economy 2019, the Department of Finance, Newfoundland & Labrador government, 2019
[p]National Energy Board, Canada, 2019
[q]Economic Report 2018, Oil & Gas UK, 2018
[r]UK Office for National Statistics, 2018
[s]Oil and Gas Production Statistics 2017–18, Scottish Government, 2019
[t]Oil and Gas Production Statistics, 2017–18, Scottish Government, 2019
[u]Energy Statistics Database, Scottish Government, 2019 (this is a provisional, unbalanced estimate figure, including Scottish oil & gas export to the rest of the world and the rest of UK)

preliminary nature of oil extraction it is not significant relative to other sectors of the economy. There are also significant proven oil reserves in Norway, Newfoundland and Labrador, and the United Kingdom (most of which are within Scotland). By contrast, volumes of proven reserves are significantly lower in Denmark, and unavailable for Iceland. In terms of ongoing extraction, Norway is the largest producer, followed by Scotland, and then by Newfoundland and Labrador. Danish oil production is significantly lower than the other cases, while Iceland is not yet engaged in oil production. In terms of oil exports, Norway is the largest exporter, followed by Scotland. Oil exports from Newfoundland and Labrador and Denmark are significantly lower than these regions, while Iceland is not yet an oil exporter. Finally, if we look at recent growth in oil production, we see that Newfoundland and Labrador is the region where production is growing, while it is declining moderately in Norway and declining more significantly in Denmark. This reflects issues of mature fields in these regions, which have been compounded by recent global price declines and volatility. Taking these indicators together, we position Newfoundland and Labrador, Norway, and Scotland as more oil oriented in terms of their political economy. By contrast, Denmark and Iceland are less oil oriented.

Norway. Following the Second World War, Norway was relatively poor compared to much of Europe, with a reliance on fisheries and agriculture as economic drivers. Norwegian oil extraction began in the North Sea in the 1970s, moved into the Norwegian Sea in the 1990s, and then moved into the Barents Sea in 2007. The Norwegian narrative of oil extraction emphasizes that the government of the day did not immediately sign over Norway's potential oil wealth to large private corporations, but demanded significant social benefits in exchange for corporate access to Norwegian oil resources (Larsen 2006). This was codified in the "Ten Oil Commandments" developed by the government to ensure employment and economic benefits for oil extraction communities and for the nation as a whole, while mitigating occupational health and safety risks, as well as environmental impacts (Urry 2013). The Oil Commandments figure prominently in the "petroleum fairy-tale" where government negotiation with the oil sector transformed the country into a place with one of the world's highest standards of living in the space of only a few decades (Kristoffersen and Dale 2014). Part of this has been the growth of the Norwegian sovereign wealth fund, where oil royalty payments to government have been reinvested in an investment fund that now ranks among the largest in the world (Bridge and Le Billon 2013; Milne 2017). This model of oil development is grounded in a political culture as a "consensus-oriented and egalitarian nation-state" (Brandal and Bratberg 2015, p. 43).

The Norwegian fairy-tale narrative also focuses on government investment and direct involvement in the oil sector through the creation of Statoil and Petoro. Government intervention "as resource-owner, regulator and entrepreneur" was built on a political culture that emphasizes that "natural resources are common property, and their use should benefit all" (Bryden 2015, pp. 146, 149). Statoil was partially privatized in 2001, though it continues to work as a hybrid public and private company that is a central actor in the Norwegian oil sector and across the North Atlantic region. In 2018, Statoil was rebranded as Equinor, reflecting a shift in the public identity of the corporation from an oil company to an energy company that is increasingly involved in renewable energy, particularly offshore wind power.

There have been few major catastrophes in the Norwegian oil sector. The most notable disaster was the 1980 Alexander L. Kailland platform sinking, which resulted in 123 deaths and is memorialized at the Norwegian Petroleum Museum in Stavanger. Through much of its history, the oil sector has been viewed positively by the public and policymakers and there has been a collaborative relationship across the oil sector and the political sphere. However, controversies have emerged around extending Norwegian oil production beyond the North Sea and Norwegian Sea operating areas. Social movements have opposed extending oil exploration into the Lofoten, Vesterålen, Senja (LoVeSe) region as well as into the Barents Sea, and Norwegian Arctic (Avango et al. 2013; Kristoffersen and Dale 2014). The push further north is a response to mature fields in the North Sea, where oil extraction volumes have been gradually declining since 1999, and the Norwegian Sea, where oil extraction volumes have been declining since 2007. At the same time, the Johan Sverdrup field, which is the largest discovery in the Norwegian Continental Shelf since the 1970s, went into production in 2019 and is expected to produce 660,000 barrels equivalent per day at its peak (Equinor 2019).

Denmark. The history of Danish oil development follows a similar trajectory as Norway. Oil extraction in the Danish Continental Shelf began in 1972. There are now 17 active oil fields, with 55 platforms. However, the "Danish shelf is both small and technically hard-to-reach" compared to neighbouring Norwegian and Scottish sites of oil extraction (Quartz + Co 2012, p. 10). While Denmark did not adopt a formal code to guide oil development like Norway's Ten Commandments, the state was involved in directing early oil development and became a stakeholder through public ownership of DONG Energy (now renamed Ørsted). Public involvement in the oil sector set expectations for social benefits of the industry and helped guide corporate social and environmental responsibility. While the Danish government still holds 58% of shares, 18% of shares were sold in 2012 to Goldman Sachs who further resold them in a 2016 IPO (Milne 2016). This caused political controversy over Goldman Sachs' profiting from the sale of shares in the public utility. A news article on the sale notes that there was significant opposition to the partial privatization of DONG energy, but that in 2012 the company was struggling and seeing declining profits. Since the change in ownership structure, it has undergone what *Financial Times* reporter Milne calls a "remarkable transformation," leveraging the capital investment from Goldman to "become the clear global leader in developing and operating offshore wind farms as it won a series of projects in the UK and elsewhere" (Milne 2016). This shift in orientation fits a longer history of Danish wind power development, which was strongly influenced by environmental movements that "formed 'critical communities' within the Danish energy sector" from the 1970s onward (Vasi 2011, p. 152).

The Danish oil sector is not as economically significant as the Norwegian sector. However, oil impacts the political economy of Denmark through taxes, trade balances, and energy security. In 2010, the oil and gas sector contributed approximately 24 billion DKK ($4.27 billion USD[1]) through corporate taxes, hydrocarbon taxes, and profit sharing

[1]Conversion rate at 2010-12-30: kr 1 DKK = $ 5.61 USD, Danmarks Nationalbank (2019).

with the government (Quartz + Co 2012, p. 12). Furthermore, the sector employs about 15,000 people, including direct and indirect employment, with most employment located on the west coast of Jutland (Quartz + Co 2012).

There have been no significant oil disasters in the Danish North Sea, which contributes to positive views of the social and environmental responsibility of the sector. Denmark has played an important role as an early adopter in offshore wind energy, which has been scaling up significantly. Oil extraction increasingly coexists with offshore wind power development in regions like Jutland. The offshore wind power transition has been facilitated by public ownership in Ørsted/DONG Energy. At the same time, volumes of oil and gas extracted from the Danish Continental Shelf are expected to decline, with increased costs for extracting the remaining oil. Strategies for dealing with decline include increasing exploration activity, increasing production in more marginal oil fields, and using technological innovation to increase the recovery rate. Declining oil extraction will result in corresponding impacts on employment and government revenues from the oil and gas sector, though some of this is offset by expanding offshore wind energy production.

In the broader context of the Kingdom of Denmark, there has been periodic interest in potential offshore oil development in Greenlandic waters since the 1970s, reflecting periods of high global oil prices and interest in Arctic oil exploration more broadly (Hansen 2014). Issues of Greenlandic oil development are bound up with issues of Danish colonialism. Many Greenlandic Inuit have been supportive of oil exploration and mining projects because resource extraction can be leveraged for greater economic and political autonomy from the Kingdom of Denmark (Shadian 2014). However, with recent declines in global oil prices and cooling interest in Arctic oil exploration an emerging Greenlandic oil economy currently seems unlikely.

Scotland. Oil development in Scotland has similarities and differences from the Nordic cases, although many in Scotland look to Norway as a model for socially beneficial oil development (Bryden et al. 2015b). The historical trajectory is similar, with an exploration and discovery phase in the 1960s and extraction ramping up through the 1970s (Bryden 2015).

Scotland produces 95% of oil within the United Kingdom and, as of 2019, is the largest oil producer in the European Union. The value of oil and gas production is estimated at £10 billion ($12.6 billion USD[2]) and supports approximately 124,500 jobs. The oil sector also contributes spin-off benefits. Oil revenues were integral developing the culture and heritage sectors on Shetland, which have become key tourism attractors (Jennings 2015).

Public investment and ownership in the Scottish oil and gas sector has been less notable than in Norway or Denmark. While publicly owned organizations like Statoil (now Equinor) and DONG Energy (now Ørsted) were core actors in the emergence and evolution of the Nordic oil sector, the Scottish sector evolved with private companies like BP and Shell as core actors. This led to a different political economy of oil. Bryden (2015) argues that UK policy was less stringent in requiring local benefits, which "led to the loss of national control and failed to generate an oil fund for future generations" (Bryden et al. 2015b, p. 17). The oil and gas sector also played into the politics of Scottish nationalism, as control over the social benefits of oil development have been part of political and public debates about the possibilities of Scottish independence (Bryden 2015).

Another key difference between Scottish and Nordic oil development is that environmental movement mobilization has previously been more visible. This is particularly the case with the 1990s controversy over Shell's proposal to sink the Brent Spar oil rig into the North Sea as a means of disposal. Greenpeace viewed the sinking of the Brent Spar as "an appropriate symbol to make the rather abstract problems of the North Sea more palpable" (Holzer 2010, p. 19). Greenpeace mobilized against the planned sinking of the platform through a public education campaign across Europe. Greenpeace also used direct action by physically occupying the oil platform (Holzer 2010). In the short term, protest and public pressure created "'reputational risk' and growing pressure from investors" on Shell to prove their commitments to environmentally responsible decommissioning practices (Barry 2015). Greenpeace

[2]Conversion rate at 2019-01-02: £1 GBP = $1.26 USD, Bank of England (2019).

successfully transformed Shell into a corporate villain in the public sphere and used public pressure to change their practices.

The Brent Spar conflict was a “critical event” (Ramos 2008) in the Scottish oil sector. Much of the Scottish North Sea is at a post-peak phase and decommissioning oil infrastructure is an issue of concern. Greenpeace’s success at making the Brent Spar a public controversy means that environmentally appropriate practices for decommissioning remain important considerations. The Scottish oil sector is dealing with mature oil fields and issues of decommissioning old oil rigs to a greater degree than our other case study regions (Hughes 2014). Scottish oil extraction peaked in 1999 and the current level of extraction “is only around 40% of the production peak in 1999” (Scottish Government 2017, p. 5). There are also moves towards renewable energy transitions by scaling up offshore wind power development. For example, in 2019, the new Beatrice Offshore Wind Farm went into operation with enough capacity to power 450,000 homes (Beatrice Offshore Windfarm 2019).

Newfoundland and Labrador. Oil development in Newfoundland and Labrador emerged later than in the North Sea, with exploration activity ramping up in the 1970s. Oil extraction has centred on three major fields: Hibernia, with the first platform going into production in 1997, as well as Terra Nova and White Rose. A fourth field, Hebron, went into production in 2017. Reserves at Hebron are estimated to be between 660 and 1005 million barrels. Hebron is projected to be in production from 2017 until 2042 (CBC 2012; Government of Newfoundland and Labrador 2012). With the later emergence and development of the oil sector, there is less concern with mature fields or decommissioning. The Newfoundland and Labrador oil industry accounts for approximately 14% of Canada’s total crude oil production (Environment Canada 2009). It has contributed significant provincial per capita GDP increases since the 1980s (Locke 2011). However, the economic benefits are less visible in smaller coastal communities than they are in the area around St. John’s, the capital city (Sinclair 2011).

The emergence of oil production in the province follows the decline of the cod fishery and the 1992 cod moratorium, arguably among the worst ecological disasters in Canadian history. The social and economic impacts were far-reaching, especially in rural coastal communities, which saw high levels of unemployment and out migration. The oil sector promised employment, economic diversification, and revitalization. For many, the oil sector is responsible for Newfoundland and Labrador's shift to "have" status from its traditional "have-not" status among Canadian provinces (CBC 2017). However, much like Scotland, Newfoundland and Labrador were less aggressive in negotiating social benefits from the energy sector. The provincial economy has also been hit by oil price declines and uncertainty in recent years. However, industry proponents maintain an optimistic tone that major new discoveries are on the horizon and that coming years will see a resurgence of the oil sector and its contributions to provincial GDP (Roberts 2019).

The oil sector has been marked by two significant workplace disasters. The 1982 sinking of the Ocean Ranger oil platform, which was carrying out exploratory work, resulted in 84 deaths. A 2009 Cougar helicopter crash, which was carrying workers to the White Rose and Hibernia oil fields, resulted in the deaths of fifteen passengers. However, Dodd argues that the "promise of oil" with its "rhetoric of job creation and Newfoundland self-determination" carries significant cultural weight in a region with a history of underdevelopment and out migration (Dodd 2012, pp. 142–143). As such, oil disasters have been reframed in public discourse as learning moments for improving workplace safety.

There have been multiple oil spills in recent years, including 250,000 litres spilled from Husky's SeaRose platform in 2018 and 12,000 litres spilled from the Hibernia platform in 2019 (CBC 2019; Kinsella 2018). These ecological harms have received media coverage, but not the kind of sustained public discussion triggered by the Ocean Ranger or Cougar helicopter disasters. By contrast, exploration proposals for hydraulic fracturing (fracking) on the west coast of the island have drawn public concern and social movement mobilization. Concerns were raised about proposed fracking in one of the enclave communities of Gros Morne National Park, a UNESCO World Heritage Site and key tourism attractor for the province (Smith 2016). In response to public concern, an expert advisory panel was convened and a moratorium was placed on further hydraulic fracturing exploratory work.

Iceland. Finally, Iceland is unique among our cases because it is not yet an oil extracting region. There are potential oil and gas resources in the Dreki area in the northeast of the country (referred to as the Dragon Field). This is close to a proposed new harbour development at Finnafjord, which would facilitate transportation to European and American markets. Exploration activity commenced in 2013, while Iceland was recovering from the 2008 economic crisis. With the impact of the financial crisis, the prospect of becoming an oil producing country was generally viewed positively in the public and political sphere. Exploration work was carried out by the Chinese oil company CNOOC (China National Offshore Oil Corporation) in collaboration with Eykon Energy, an Icelandic company. An exploration license was also held by Norway's Petoro. Exploratory work did not provide a sense of whether there are suitable oil and gas resources to proceed with extraction. With a global downturn in oil prices, Arctic oil exploration programmes have become less viable. Prior to the price downturn, there was political consensus that Icelandic oil was worth pursuing as an economic development strategy. The hope was for a Nordic model of development, following the leads of Norway and Denmark. However, unless oil prices rebound significantly, debate over how to distribute Iceland's oil resources is likely to remain peripheral.

After a long cycle of high-oil prices through the early 2000s, there was a major drop in global oil prices in 2014, followed by ongoing volatility in the global oil economy. This has been driven by events outside the North Atlantic. The rapid increase in hydraulic fracturing in the United States has seen the U.S. regain its status as a major oil source region. This fracking boom, coupled with the responses of OPEC, has contributed significantly to the recent period of volatility, which led many companies to restructure their operations in the interests of economic efficiency. At the time of writing, this volatility is further compounded by the disruptions of the COVID-19 health pandemic, as well as tensions between OPEC and Russia. This ongoing period of price volatility creates pressure on host societies to revisit regulatory and royalty regimes under the rhetoric of global competitiveness. It also dampened exploratory activity, particularly in the challenging and costly operating environments of the Arctic and northern reaches of the North Atlantic, which had a cooling effect in Iceland. CNOOC and Petoro let go of their exploration licenses, and political and public debate about oil development has largely fallen beneath the surface (Ćirić 2018).

Tourism Development in the North Atlantic

Nature-based tourism involving hunting and fishing has been established across the North Atlantic region for a long time. However, recent decades have seen a turn towards models of eco-tourism, sustainable tourism, or geo-tourism as alternate ways of pursuing lower-impact forms of nature-based tourism (Urry and Larsen 2011). Eco-tourism is premised on non-extractive interaction with natural environments through activities like hiking, sea kayaking, surfing, cycling, or birdwatching, rather than extractive recreation like hunting and fishing. Increasing interest in travel to rural, remote, and cold-water regions has been a parallel trend, as travellers seek alternatives to mass market, resort-based tourism (George et al. 2009). Communities across the North Atlantic have turned to tourism as a means of economic diversification and the revitalization of rural coastal communities (Baldacchino 2010; Huijbens and Jóhannesson 2019). While tourism is often viewed positively, it also creates tensions with established livelihoods and community identities, for example those based on fishing (Ounanian 2019).

Table 2.2 presents indicators of the study regions in terms of tourism-orientation. If we look at tourism employment as a percentage of the population, then Iceland (18.79%) has a very high level of tourism employment. This is followed by Norway (6.36%), Scotland (3.8%), Denmark (3.97%), and then Newfoundland and Labrador (3.42%). However, if we look at inbound tourism traffic, we see a different picture. Denmark and Scotland have the largest visitor flows by far, followed by Norway, then Iceland, then Newfoundland and Labrador. To put this in a different context, we can look at inbound tourism numbers as a ratio of the host community population, which has been used to measure overtourism (Oklevik et al. 2019). While it is difficult to compare country-level and city-level figures, Iceland and Denmark stand out with ratios that are similar to well-known European city destinations like Barcelona (1: 6.1), though their ratios are substantially lower

Table 2.2 Tourism orientations of case study regions

	Tourism Sector Direct Employment (2018, unless otherwise noted)[a]	Population (2018 unless otherwise noted)	Tourism sector employment as % of population	Inbound tourism: number of tourist arrivals per year (2017, unless otherwise noted)[b]	Ratio of resident population to inbound tourists per year
Denmark	229,600	5,781,190[c]	3.971	29,730,000	1: 5.14
Iceland	67,100	357,050[d]	18.792	2,353,000	1: 6.59
Norway	336,700	5,295,619[e]	6.358	6,252,000[f]	1: 1.18
Newfoundland and Labrador	18,000[g]	525,075[h]	3.428	510,876[i]	1: 0.97
Scotland	207, 000[j] (2016)	5,404,700[k] (2016)	3.830	14,300,000[l] (2016)	1: 2.64

[a]World Travel & Tourism Council, 2019
[b]United Nations World Tourism Organization, 2019
[c]Statistics Denmark, 2019
[d]Statistics Iceland, 2018
[e]Statistics Norway, 2019
[f]Data does not include same-day visitors (excursionists)
[g]The Department of Tourism, Culture, Industry and Innovation Government, Newfoundland and Labrador Government, 2019
[h]Statistics Canada, 2019
[i]The Department of Tourism, Culture, Industry and Innovation Government, Newfoundland and Labrador Government, 2019
[j]Scottish Government, 2017
[k]UK Office for National Statistics, 2017
[l]Scottish Government, 2017

than the most overtouristed European destinations like Venice (1: 16.3) or Amsterdam (1: 11.5). These cases are followed by Scotland, then Norway, then Newfoundland and Labrador. While each of these indicators paints a different picture, taking them together we characterize Scotland, Iceland, and Denmark as highly tourism-oriented case study regions. By contrast, Newfoundland and Labrador and Norway are relatively less tourism-oriented cases, with tourism being important to particular regions or local communities.

Scotland. Among our case study regions, Scotland has a long history as a destination for nature-oriented tourists in the United Kingdom. The tourism industry has long had an influence on "narratives of Scottishness" through its global circulation of imagery (Edensor 2002). Early forms of nature-based tourism oriented around sport activities like hunting and fishing. The tourism sector in Scotland evolved and diversified over time. It is now "mature" and well-developed, with multiple foci that target visitors with interests in historic sites and museums, arts and culture, food and drink (e.g. whiskey tourism), and nature experiences and outdoor recreation (VisitScotland 2014). The tourism sector evolved with a dual focus on rural environments and urban centres, particularly Edinburgh. Tourism also plays a key role in the political economy of Scotland, with 217,000 tourism jobs accounting for 9% of total employment in 2014 (Visit Scotland 2016). Beyond the direct economic impacts, tourism development has promoted the growth of heritage and cultural sectors in rural and remote communities, such as Shetland, which is characterized by an increasingly "vibrant cultural scene" (Jennings 2015, p. 173). While tourism is important to the Scottish political economy, the uncertainties of Brexit are causing concern, particularly as the EU is a core source market for Scottish tourism and EU nationals make up a significant proportion of the tourism sector workforce (Scottish Government 2018).

Newfoundland and Labrador. Nature-oriented tourism in Newfoundland and Labrador emerged as early as the 1890s, with a focus on camping and sport fishing. During the 1920s and 1930s, nature-based tourism was oriented around sport fishing and hunting, with camps and lodges as destinations. Visiting photographers and writers popularized the region through images of outport villages, rocky shores and headlands, and icebergs—an iconography that still circulates in tourism marketing. This early wave of tourism development abated following World War Two. However, there was a resurgence of interest in tourism development after the 1992 cod fishing moratorium (Ommer 2007). Tourism development initially focused on the history and perceived authenticity of rural "outport" fishing communities. Over time, tourism developed a dual focus on history and culture, on one hand, and experiential interactions with nature through hiking and boat tours, on the other hand. Icebergs, puffins and whales are particularly iconic nature-based tourism attractors (Stoddart and Graham 2016; Stoddart and Sodero 2015).

While not as economically significant as the oil sector, the tourism sector has grown significantly over time in terms of revenues and employment. In 1992, approximately 264,000 non-residents visited the province. By 2010, this number increased to 518,500 visitors and an estimated $411 million Canadian dollars ($402.9 million USD[3]) in tourism revenues (Globe 2011). The success of tourism marketing is reflected in the province's designation as "one of the top 'new' and 'undiscovered' travel destinations on … lists from industry leaders like Lonely Planet and Fodor's" (Government of Newfoundland and Labrador 2009, p. 18). Tourism is seen as an important avenue for economic diversification, especially for rural coastal communities beyond the capital city of St. John's. For example, Gros Morne National Park, which is a UNESCO World Heritage Site, is an important nature-based tourism hub. Gros Morne was established in the early 1970s and was intended to bring "the maximum funds into the local economy" (Overton 1996, p. 182). Currently, Gros Morne brings in "upwards of 180,000 visitors each year … accounting for in excess of $37 million [Canadian dollars] in economic benefits" ($33.6 million USD[4]) (Parks Canada, as cited in Canada-Newfoundland and Labrador Offshore Petroleum Board 2014, p. 394). As we explore in our analysis, this region has also been at the centre of controversies around oil development.

[3]Conversion rate at 2010-01-01: $1 CAD = $0.98 USD, Bank of Canada (2019).

[4]Conversion rate at 2014-01-01: $1 CAD = $0.90 USD, Bank of Canada (2019).

Iceland. By contrast, Iceland has a more recent and idiosyncratic history of tourism development. Until recently, the country was a niche destination for adventurous travellers interested in outdoor recreation. However, Iceland experienced a recent tourism boom and tourism is "hailed as one of three pillars of the Icelandic economy, together with fisheries and heavy industries" (Huijbens and Jóhannesson 2019, pp. 281–282). This can be traced to the unintended consequences of the Eyjafjallajökull volcano eruption in 2010 (Benediktsson et al. 2011). This event grounded flights across Europe and garnered international media attention. Imagery of the dramatic Icelandic landscape spread through media coverage and generated interest, which was leveraged by Icelandair and the tourism sector for promotion to North American and European markets (Lund et al. 2017). Since the eruption, annual growth in tourism visitation has been dramatic. The declining value of the Icelandic Krona (ISK) in the wake of the 2008 global economic crisis also helped push tourism development as a recovery strategy (Antanova and Rieser 2019; Lund et al. 2018).

Tourism quickly became significant to the Icelandic political economy. Tourism revenues nearly doubled between 2012 and 2016 to 466,287 billion Icelandic Krona ($4131 billion USD[5]). As of 2016, tourism revenues account for 39.2% of economic activity, which surpasses established industries including "exports of marine products and industrial products" (Óladóttir 2017, p. 2). The number of tourism employees increased from 19,500 to 28,900 between 2012 and 2016, while the number of international visitors went from 488,600 to 1,791,400 annually during the same period (Óladóttir 2017, p. 7). As Lund et al. (2017) note, "Iceland is now a widely popular world destination – ranking 16th and 11th on the World Economic Forum's 2013 and 2011 Travel and Tourism Competitiveness Reports, respectively" (Lund et al. 2017, pp. 149–150). Tourism development is also seen as particularly important for promoting the well-being of rural areas, as it may serve as a collaboration and innovation space among private businesses, the public sector, and academic and research institutions (Edvardsdóttir 2016). However, 2019 figures from Statistics Iceland show declining visitor

[5]Conversion rate at 2016-12-30: kr 1 ISK = $0.00886 USD, Central Bank of Iceland (2019).

numbers, which suggests the tourism boom may be softening (Statistics Iceland 2019). This raises questions about the long-term economic and social sustainability of tourism as a development pathway.

Icelandic tourism is oriented around nature-oriented experiences, such as whale and puffin watching, the northern lights, glaciers, hiking, and geothermal pools, such as the iconic Blue Lagoon (Huijbens 2016). Tourism promotion focuses on images of Arctic landscapes and nature, thereby playing on increased interest in Arctic tourism more broadly. At the same time, the country is defined as being both exotically Arctic but familiarly European, with the capital city of Reykjavik providing "museums, shopping, contemporary design, art, and a lively nightlife" (Lund et al. 2018, p. 144). There are, however tensions between ongoing whale hunting and the increasing popularity of whale watching, which has been positioned as an environmentally responsible alternative to whale hunting (Cunningham et al. 2012; Hall et al. 2009). Tourism companies like North Sailing and Gentle Giants have played a particularly important role in transforming the small community of Húsavík, on the north coast of the country, into a popular hub for whale tourism, thereby "rejuvenating the local economy" (Cunningham et al. 2012, p. 151). This provides an example of how extractive economies can be transformed through nature-oriented tourism in ways that contribute to the economic revitalization of rural coastal communities.

Denmark. Denmark has also seen a recent expansion of its tourism economy. Much tourism traffic focuses on the capital city of Copenhagen, which has become a trendy urban destination where tourism attractors include history, arts, and culture, and urban environments like the harbour. Though Copenhagen is an urban tourism destination, it can also be seen as an eco-tourism site because cycling is integral to the tourism destination image and place-making of the city (Nilsson 2019). Elsewhere, rural tourism orients around beaches and coastlines (Baerenholdt et al. 2004). There is a parallel focus on local craft as a form of rural tourism entrepreneurship in places like the island of Bornholm (Prince 2017). Like Iceland, tourism has grown over the past several years, though the pace of growth is less dramatic. Visitation numbers have grown 16.9% between 2012 and 2016, with current trends suggesting the country may receive more than 22 million tourists by 2020 (Statistics Denmark 2017; World Tourism Organization 2017). According to VisitDenmark (2016), tourism is also a major employment driver, with over 100,000 direct and indirect jobs in the sector.

In the broader context of the Kingdom of Denmark, Greenland is emerging as a niche destination as increasing numbers of travellers want to witness real-time effects of climate change. Greenlandic tourism is built on both the promise of encountering "serene, majestic" nature and the promise of the "last chance to see" endangered forms of Arctic nature and wildlife (Ren et al. 2016, p. 289). Tourism is positioned as an economic pillar for the future of Greenland and as one development pathway (along with mining and fisheries) that can be leveraged for greater autonomy from Denmark. However, as a tourism host community, Greenland faces challenges of remoteness and accessibility to tourism markets, as well as "difficulty of attracting investment, high seasonality and highly sensitive environments" (Ren et al. 2016, p. 285).

Norway. In contrast to the other cases, Norway's tourism sector is less developed. Part of the explanation is the high cost of travel to and within Norway. Nevertheless, natural environments and outdoor recreation are important attractors for Norwegian tourism. The coastal fjord landscape of the west coast is an important tourism region that shows signs of overtourism (Oklevik et al. 2019). The Norwegian Arctic is increasingly the site of tourism travel, motivated by experiencing Europe's "northern extremity" and the impulse to witness Arctic environments as they are radically transformed by climate change (Jacobsen 2015, p. 136). Indigenous tourism focused on traditional Sami culture is also a growing area (Olsen 2016).

The Lofoten Islands region, above the Arctic Circle where the Norwegian Sea meets the Barents Sea, is going through a localized tourism boom. There are about 180 tourism companies operating in the area "with an estimated local annual value generation of 40 million euros" (Kaltenborn et al. 2017, p. 32). Estimates are that about a million tourists visit Lofoten per year, a rural region that is "home to just 23,000 residents" (Milne 2017). The growing popularity of Lofoten as a tourism hub reflects broader trends in increasing Arctic tourism travel but is also amplified by pop culture imagery. Lofoten embodies the fictionalized northern setting of Disney's *Frozen* films, which is a connection regularly made in news stories about the tourism boom. Tourism is creating tensions within host communities around how tourism development should be managed, whether as an engine for economic development or for building local community wellbeing. This raises questions of whether (or how) to intervene in tourism development to discourage mass tourism, prioritize local initiatives, and protect the natural environment and heritage of the region (Lindberg et al. 2019). One response is a national promotion strategy that encourages diffusion away from the major hubs with the messaging "all of Norway, all year round" (VisitNorway 2019).

While tourism is secondary to the national Norwegian political economy, there are regional hubs where it is significant and growing. Lofoten is also significant as a place of contact and controversy between oil development and tourism development. As Kristoffersen notes, "Lofoten is the foremost symbol and icon in national politics regarding whether or not commercial drilling should be allowed in the region, due to its key role historically and culturally, and as a tourist attraction, potentially becoming a World Heritage Site" (Kristoffersen 2014, p. 8). Lofoten is an important site of contact between tourism and oil development that we explore in the remainder of the book.

Responses to Climate Change in the North Atlantic

As noted in the previous chapter, global climate change is an important meta-context of the oil-tourism interface. Ongoing and expected ecological impacts of climate change across the North Atlantic will affect the political economy of oil extraction and eco-tourism. Melting sea ice is opening northern frontier regions for potential natural resource exploration in Norway, Iceland, and Greenland (Bridge and le Billon 2013; Shadian 2014). In Newfoundland and Labrador, changes to sea ice lead some to predict fewer lost operating days in the oil sector due to sea ice and icebergs. Climate change is already having significant ecological impacts on northern and Arctic landscapes, including shifting vegetation ranges and wildlife habitat, melting sea ice and permafrost (Ren et al. 2016). Sea level rise, coastal erosion, flooding, and increasingly severe storms are also among the regional impacts of climate change (Wejs et al. 2014). Furthermore, global social and political responses to climate change have the capacity to reconfigure tourism practices and energy systems in the coming decades, with potentially significant impacts for host communities that rely on these development pathways. As such, it is worth considering how our case study regions are responding to the challenges of climate change.

Table 2.3 presents indicators of climate change policy and performance for comparison of these five regions. Iceland has the largest share of renewable energy as a percentage of final energy consumption, while also having low per capita carbon emissions, though its most

Table 2.3 Climate policy and performance orientations of case study regions

	Renewable sources, % of total final energy consumption (2015)[a]	Carbon dioxide emissions per capita in tonnes (2014)[b]	Climate Change Performance Index (CCPI), Overall Performance Index (2019)[c]	Climate Change Performance Index (CCPI), Ranking (2019)[d]	Climate Change Performance Index (CCPI), 10-year Ranking Trend (2010–2019)[e]
Denmark	33.2	5.9	61.96	15	
Iceland	77.0	6.1	52.55 (2017)[f]	39 (2017)	
Norway	57.8	9.3	62.80	12	
Newfoundland and Labrador (Canada)	22.0 (Canada)	15.1 (Canada)	34.26 (Canada)	54	
Scotland (UK)	8.7 (UK)	6.5 (UK)	65.92 (UK)	8	

[a]World Development Indicators database, World Bank, 2019
[b]World Development Indicators database, World Bank, 2019
[c]Climate Change Performance Index, 2019
[d]Climate Change Performance Index, 2019
[e]Climate Change Performance Index, 2019
[f]For methodological reasons, Iceland is not included in the 2019 CCPI

recent international Climate Change Performance Index (CCPI) ranking is 39. The United Kingdom, Norway, and Denmark rank highly on the CCPI, with relatively low per capita carbon emissions among the five regions. The national UK ranking masks regional variation, with Scotland often serving as a climate leader within the United Kingdom. Due to recent policy shifts, Denmark dropped in the climate change performance index rankings, whereas it had previously been among the top-ranked countries. While wind power is scaling up in Denmark, renewable energy development, as a percentage of energy consumption, lags beyond Iceland and Norway. Finally, Canada continues to rank low on the CCPI, with the highest per capita carbon footprint, as well as a lower percentage of renewable energy than the other regions except the United Kingdom. The Canadian ranking also masks regional variation. Newfoundland and Labrador is a laggard relative to other provinces with proactive climate change orientations, such as British Columbia or Quebec. Taking these measures together, the main divide is between Newfoundland and Labrador (Canada), as a poor performer, versus the other four case study regions.

Denmark. Environmental concerns were integrated into policy "unusually fast in comparison with other countries," partly in response to early waves of mobilization by the Danish environmental movement (Wettergren and Soneryd 2017, p. 171; also see Vasi 2011). Denmark ranks substantially higher than other European countries on measures of environmental activism, particularly around petitioning and contributing to environmental organizations (Bozonnet 2017). The "greening" of the state is supported by high levels of public concern and political consensus on the need for climate policy, including the need to connect climate policy and domestic energy security. Lines of political debate focus on the "prominence of state involvement" and reflect urban–rural tensions in the country (Ladrech and Little 2019, p. 1024). Building on energy efficiency measures and renewable energy development, Denmark cut its CO_2 emissions by 32.6% between 1990 and 2014 (Danish Energy Agency 2014). During the same period, energy consumption in Denmark declined by 7.8%. After leading the global CCPI index for five years, it dropped to thirteenth place, which reflects the shifting climate policy orientations of the centre-right coalition government that came to office in 2015. However, in terms of "emissions development, renewable energy and energy efficiency the country still performs within the top group" (Burck et al. 2017, p. 4). The 2019 Danish election was viewed as a "climate change election," with a majority of Danes seeing climate change as a particularly significant issue (Henley 2019). Mette Frederiksen's Social Democratic government set GHG reductions targets of 70% by 2030, which goes further than targets for the European Union as whole.

Denmark took an early leadership role in developing wind energy, which is "an important part of the 'branding' of Denmark as 'world-leading' in ecological modernization" (Wettergren and Soneryd 2017, p. 174). Renewable energy production doubled between 1990 and 2014 and now accounts for over 25% of total energy consumption, but over 50% of energy used for heating (Danish Energy Agency 2014). Esbjerg, which is a hub city for the oil sector, has become an important entry point for North Sea offshore wind installation. The Danish image as a green country committed to climate action and renewable energy transitions is a reason why the oil sector seems marginal in public and political discourse (Sovacool et al. 2017; Vasi 2011).

Denmark was the site of the 2009 COP 15 (Conference of the Parties) international climate change meetings. This event provoked one of the largest episodes of environmental movement mobilization around the COP meetings. Movement actors emphasized climate justice discourse, while anti-capitalist activists used digital media to critique the COP process as neoliberal greenwashing (Cassegård and Thörn 2017; Hadden 2015; Uldam and Askanius 2013). The meetings failed to produce substantial progress on international climate governance and were interpreted as a defeat in the Danish environmental movement. The aftermath of COP 15 changed the environmental movement in Denmark. Institutionalized national organizations and chapters of international organizations remained active, while smaller, grassroots, and radical groups went into abatement. As a result, Danish environmental politics saw a retrenchment to institutionalized political spheres and in-system tactics after the COP 15 mobilization (Cassegård and Thörn 2017). However, there have been recent episodes of environmental movement mobilization, such as fossil fuel divestment and anti-fracking movements, which "call Denmark into question as a green nation, and that point to its fossil fuel dependence" (Wettergren and Soneryd 2017, p. 185).

Norway. While Denmark projects a self-image as a pro-environmental society that has embarked on the renewable energy transition, Norway is more complicated. Like Denmark, Norway ranks substantially higher than other European countries on measures of environmentalism, particularly around petitioning and contributing to environmental organizations (Bozonnet 2017). The Norwegian government proclaims its commitment to the Paris agreement goals, pursues "'disruptive' international initiatives such as divestment from coal in its pension fund," and facilitates the electrification of car transportation (Ćetković and Skjærseth 2019, p. 1054). However, the oil sector is a core part of the political economy and government remains committed to oil exploration and the expansion of frontier areas in the "High North," including into the Barents Sea and Norwegian Arctic (Kristoffersen and Dale 2014). At the same time, the Norwegian government and oil sector are scaling up investment in offshore wind development. As part of this shift, Statoil rebranded as Equinor to signal its transformation from an oil company to a more holistic energy company. A corporatist political culture with well-established relationships between government, environmental groups and fossil fuel companies has helped build a "cross-party consensus on climate change" and encouraged transitions in the energy sector (Farstad 2019, p. 1010).

Norgaard's (2011) research captures this multifaceted, often contradictory, stance regarding climate change and oil production in Norway. She shows how everyday talk in the political and public sphere works to legitimate continuing oil development, while also acknowledging the severity of climate change. Common themes include the notion that Norway is a small country in a global system of carbon emissions and that Norway suffered a long history as an underdeveloped part of Europe prior to its oil wealth. Furthermore, there is a discourse that Norwegian oil is produced to higher environmental and social standards than oil from other countries and it is beneficial to export it into the global energy system because it can displace dirtier fossil fuels. These discourses allow Norwegians to maintain an everyday form of climate denial that masks the tensions between espousing concern about climate change, committing to low-carbon climate policies, and continuing fossil fuel extraction.

Iceland. Iceland presents the self-image of a green society with well-developed renewable energy infrastructure based on geothermal and hydropower. Most household energy use comes from renewable energy sources. However, carbon footprints for transportation and shipping are more significant, which reflects the country's small and diffuse population. Tourism-related air travel is also a significant contributor to the country's carbon footprint (Huijbens 2016). However, the tourism sector's carbon footprint gets limited attention in public or political discourse about responding to climate change, or in discussions of landscape transformations that are already occurring, most notably melting glaciers. As oil extraction off the northeast coast is still hypothetical, the fossil fuel sector does not figure into Iceland's commitments to climate change policy and action. Instead, there are tensions between government, the energy sector, and environmental organizations around the impacts of new renewable energy infrastructure development, which potentially infringes on wilderness values that draw tourists to the country.

Scotland. Like Denmark and Norway, the United Kingdom ranks higher than average on measures of environmental participation, particularly on petitioning and contributing to environmental organizations (Bozonnet 2017). Scotland is a climate leader within the United Kingdom, with the SNP government's Climate Change Bill setting targets to reduce or offset all GHG emissions by 2045 (BBC 2019a). A key policy is the declaration of a ban on gas and diesel cars with a 2032 target. Looking at the United Kingdom as a whole, however, the 2018 CCPI report notes that "long-term 2030 targets for emissions and renewable energy are not ambitious enough for a well-below-2°C pathway" (Burck et al. 2018, p. 4). As an oil-dependent region, Scotland is dealing with mature, post-peak oil fields in the Scottish North Sea, as well as decommissioning rigs at the end of their production cycles. Relatedly, there is talk about using depleted oil fields for carbon capture and storage to remove CO_2 from the atmosphere. Technologies for $C0_2$ Utilization (CCU) are also being explored, which would allow CO_2

to be "recovered and converted into commercially valuable products, including bio-oils, chemicals, fertilisers and fuels" (Scottish Government 2017, p. 35). There are also moves to rapidly scale up offshore wind production as part of a renewable energy transition. In 2019, for example, wind power capacity reached a milestone as more than enough power was produced to meet the needs for household energy consumption (BBC 2019b). As with Denmark, climate mitigation strategies are framed in terms of reducing energy consumption and carbon footprints, but also framed in terms of creating new opportunities through emerging economies and technological innovation (Oxburgh 2016).

Newfoundland and Labrador. Finally, Newfoundland and Labrador is among the Canadian provinces that are reluctant to adopt ambitious climate mitigation policies, or to support federal government efforts to reduce Canada's climate impacts. Provincial climate change policy focuses on individual-level education and supporting community-level adaptation to changing climatic conditions. There is less evidence of climate policy that challenge the oil sector's political economic power or that puts new investment at risk (Sodero and Stoddart 2015). In their analysis of Canadian oil sector, Carroll et al. (2018) talk about the shift from stage 1 to stage 2 forms of climate denial in the oil sector. Stage 1 denial means questioning the reality or significance of climate change, which is a stance no longer held by most of the oil sector. Instead, stage 2 denial focuses on promoting government "policies that appear as credible responses to the scientific consensus but do not harm big carbon" (Carroll et al. 2018, p. 428). The focus of Newfoundland and Labrador on education and community adaptation, which avoids challenging the political economic power of the oil sector, is a form of stage 2 denial. Furthermore, the province is the least developed among our study regions from a renewable energy transition perspective. A central part of the provincial plan to address climate change is to replace oil-based electricity generation with large-scale hydroelectric power from the Muskrat Falls project. This project is contentious because of massive cost overruns and delays. It has also been the subject of mobilization by Inuit and NunatuKavut Indigenous land protectors and their allies. Environmental justice concerns include downstream impacts of methylmercury on fish populations, which are integral to local diets, as well as fears about infrastructure failure that could result in community flooding and displacement (Allen 2017).

Summary of the Case Studies

This chapter described facets of the political economies of our case study regions to contextualize the next three chapters on cultural dimensions, governance, and social movement strategic interaction in the oil-tourism interface. These political economy orientations speak to the relative economic power of the oil and tourism sectors across our case studies, which can translate into cultural visibility and political efficacy. Figure 2.1 visualizes our case study regions according to their oil and tourism orientations. This figure represents the relative similarities and differences among our cases as a model with two dimensions. The horizontal dimension represents low- to high-tourism orientation, and the vertical dimension represents low- to high-oil orientation. Our cases are positioned relative to one another in the resulting grid.

There are significant similarities and differences that allow for analytically meaningful comparisons across regions. In visualizing the cases this way, we locate Scotland as high-oil and high-tourism in the orientation of its political economy in relation to the other cases. Denmark and Iceland, by contrast, have relatively low-oil and high-tourism orientations in relation to the other cases. Finally, Newfoundland and Labrador and Norway have relatively high-oil and low-tourism orientations in relation to the other cases. As a qualification, our analysis excludes negative cases, or regions that have low-oil and low-tourism orientations. A comparison with negative cases is a valuable avenue for further research. With the political economic context of the different cases in mind, the next three chapters engage in a comparative analysis of the cultural, governance, and social movements dimensions of the oil-tourism interface in the North Atlantic.

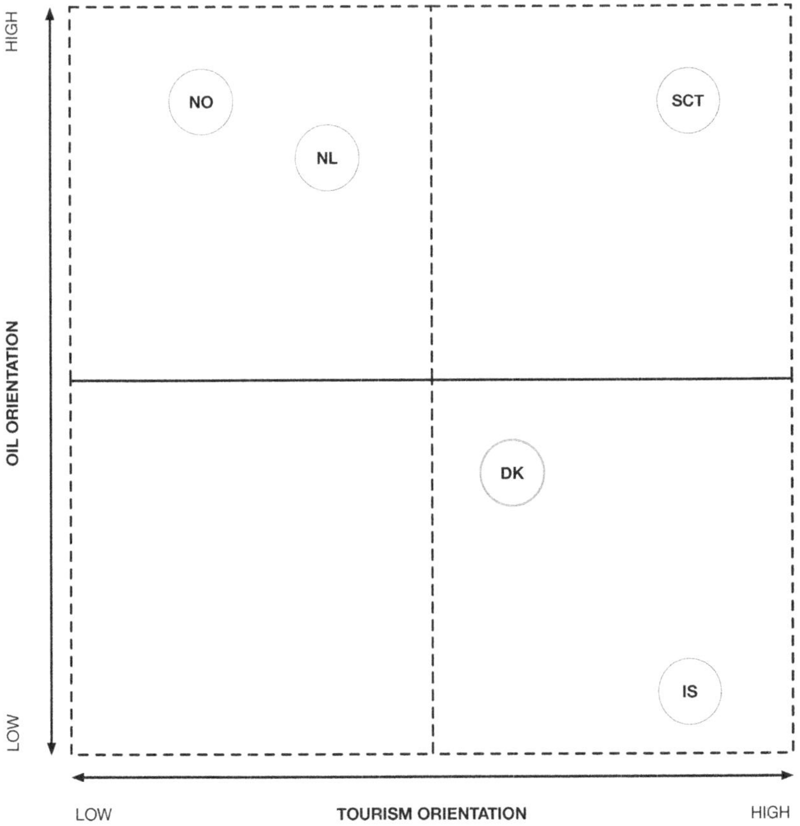

Fig. 2.1 Synthesis of oil and tourism orientations of case study regions

References

Allen, V. (2017). Muskrat Falls. In A. Marland & L. Moore (Eds.), *The democracy cookbook: Recipes to renew governance in Newfoundland and Labrador* (pp. 318–320). St. John's, NL: ISER Books.

Avango, D., Nilsson, A. E., & Roberts, P. (2013). Assessing Arctic futures: Voices, resources and governance. *The Polar Journal, 3*(2), 431–446.

Baerenholdt, J. O., Haldrup, M., Larsen, J., & Urry, J. (2004). *Performing tourist places*. Aldershot, UK: Ashgate.

Baldacchino, G. (2010). *Island enclaves: Offshoring strategies, creative governance, and subnational island jurisdictions*. Montreal and Kingston: McGill-Queen's Press.

Baldacchino, G., Greenwood, R., & Felt, L. (Eds.). (2009). *Remote control: Governance lessons for and from small, insular, and remote regions*. St. John's, NL: ISER Books.

Baldacchino, G., Vodden, K., & Gibson, R. (Eds.). (2015). *Place peripheral: Place-based development in rural, island and remote regions*. St. John's, NL: ISER Books.

Barry, A. (2015). The oil archives. In A. Hannah, M. Arthu, & M. Watts (Eds.), *Subterranean estates: Life worlds of oil and gas* (pp. 95–107). Ithaca, NY: Cornell University Press.

BBC. (2019a). *Climate change: MSPs approve beefed up emissions target*. https://www.bbc.com/news/uk-scotland-49819905. Accessed 28 January 2020.

BBC. (2019b). *Scottish wind power output at record high*. https://www.bbc.com/news/uk-scotland-scotland-business-48981626. Accessed 28 January 2020.

Beatrice Offshore Windfarm. (2019). *Scotland's largest operational offshore windfarm*. https://www.beatricewind.com/. Accessed 28 January 2020.

Beck, U., & Levy, D. (2013). Cosmopolitanized nations: Re-imagining collectivity in world risk society. *Theory, Culture & Society, 30*(2), 3–31.

Benediktsson, K., Lund, K. A., & Huijbens, E. (2011). Inspired by eruptions? Eyjafjallajökull and Icelandic tourism. *Mobilities, 6*(1), 77–84.

Bozonnet, J.-P. (2017). Explaining environmental activism by national cultures: The hypothesis of hysteresis. In A. Telesiene & M. Gross (Eds.), *Green European: Environmental behaviour and attitudes in Europe in a historical and cross-cultural comparative perspective* (pp. 91–110). London: Routledge.

Brandal, N., & Bratberg, Ø. (2015). Cousins divided? Development in and of political institutions in Scotland and Norway since 1814. In J. Bryden, O. Brox, & L. Riddoch (Eds.), *Northern neighbours: Scotland and Norway since 1800* (pp. 42–62). Edinburgh, UK: Edinburgh University Press.

Bridge, G., & le Billon, P. (2013). *Oil*. Cambridge, UK: Polity.

Brinklow, L., & Gibson, R. (Eds.). (2017). *From black horses to white steeds: Building community resilience*. Charlottetown, PEI: Island Studies Press at the University of Prince Edward Island.

Bryden, J. (2015). The development of industry and North Sea oil in Scotland and Norway. In J. Bryden, O. Brox, & L. Riddoch (Eds.), *Northern neighbours: Scotland and Norway since 1800* (pp. 123–153). Edinburgh, UK: Edinburgh University Press.

Bryden, J., Brox, O., & Riddoch, L. (Eds.) (2015a). *Northern neighbours: Scotland and Norway since 1800*. Edinburgh, UK: Edinburgh University Press.

Bryden, J., Opsahl, E., Brox, O., & Riddoch, L. (2015b). Introduction. In J. Bryden, O. Brox, & L. Riddoch (Eds.), *Northern neighbours: Scotland and Norway since 1800* (pp. 1–26). Edinburgh, UK: Edinburgh University Press.

Burck, J., Marten, F., Bals, C., Höhne, N., Frisch, C., Clement, N., et al. (2018). *The climate change performance index: Results 2018*. Bonn, Germany: Germanwatch, Climate Action Network International and NewClimate Institute.

Burck, J., Marten, F., Bals, C., Uhlich, T., Höhne, N., Gonzales, S., et al. (2017). *Climate Change Performance Index 2017*. Bonn, Germany: Germanwatch Nord-Süd Initiative.

Canada-Newfoundland and Labrador Offshore Petroleum Board. (2014). *Western Newfoundland and Labrador offshore area Strategic Environmental Assessment update: Final report*. St. John's, NL: Canada-Newfoundland and Labrador Offshore Petroleum Board.

Carroll, W., Graham, N., Lang, M. K., Yunker, Z., & McCartney, K. D. (2018). The corporate elite and the architecture of climate change denial: A network analysis of carbon capital's reach into civil society. *Canadian Review of Sociology, 55*(3), 425–450.

Cassegård, C., & Thörn, H. (2017). Climate justice, equity and movement mobilization. In C. Cassegard, L. Soneryd, H. Thorn, & A. Wettergren (Eds.), *Climate action in a globalizing world: Comparative perspectives on environmental movements in the global north* (pp. 33–56). New York: Routledge.

CBC. (2012). *Hebron megaproject gets qualified endorsement: Helicopter safety emphasized in report; doubt given to labour projections*. CBC. http://www.cbc.ca/news/business/story/2012/02/28/nl-hebron-review-commission-ayre-228.html?cmp=rss. Accessed 2 July 2014.

CBC. (2017). *Hibernia celebrates 20 years since first oil*. CBC. http://www.cbc.ca/news/canada/newfoundland-labrador/20th-anniversary-of-hibernia-1.4406683. Accessed 19 November 2017.

CBC. (2019). *12,000 L of oil spilled into ocean off Newfoundland, causing oil rig shutdown*. https://www.cbc.ca/news/canada/newfoundland-labrador/hibernia-oil-spill-production-stopped-1.5216108. Accessed 27 January 2020.

Ćetković, S., & Skjærseth, J. B. (2019). Creative and disruptive elements in Norway's climate policy mix: The small-state perspective. *Environmental Politics, 28*(6), 1039–1060.

Ćirić, J. (2018). Two companies halt oil exploration in Icelandic waters. *Iceland Review*. http://icelandreview.com/news/2018/01/23/two-companies-halt-oil-exploration-icelandic-waters. Accessed 15 August 2018.

Cunningham, P. A., Huijbens, E. H., & Wearing, S. L. (2012). From whaling to whale watching: Examining sustainability and cultural rhetoric. *Journal of Sustainable Tourism, 20*(1), 143–161.

Danish Energy Agency. (2014). *Energy statistics 2014: Data, tables, statistics and maps*. Copenhagen, Denmark: Danish Energy Agency.

Dodd, S. (2012). *The Ocean Ranger: Remaking the promise of oil*. Halifax, NS: Fernwood.

Edensor, T. (2002). *National identity, popular culture and everyday life*. Oxford, UK: Berg.

Edvardsdóttir, A. G. (2016). *The interaction of the knowledge society and rural development in Iceland and Scotland*. (PhD). University of Iceland, Reykjavik, Iceland.

Environment Canada. (2009). *National inventory report 1990–2007: Greenhouse gas sources and sinks in Canada*. http://www.ec.gc.ca/Publications/B77E6264-D0E3-45B5-BE56-5A395C0B7653%5CNationalInventoryReport19902007GreenhouseGasSourcesAndSinksInCanada.pdf. Accessed 5 March 2012.

Equinor. (2019). *Johan Sverdrup is now open. That's good news for our investors, for Norway—and for emissions*. https://www.equinor.com/en/what-we-do/johan-sverdrup.html. Accessed 28 January 2020.

Farstad, F. M. (2019). Does size matter? Comparing the party politics of climate change in Australia and Norway. *Environmental Politics, 28*(6), 997–1016.

Felt, L. (2009). A tale of two towns: Municipal agency and socio-economic development in Akureyri, Iceland, and Corner Brook, Newfoundland. In G. Baldacchino, R. Greenwood, & L. Felt (Eds.), *Remote control: Governance lessons for and from small, insular, and remote regions* (pp. 148–169). St. John's, NL: ISER Books.

Freudenburg, W. R., & Gramling, R. (2011). *Blowout in the Gulf: The BP oil spill disaster and the future of energy in America*. Cambridge, MA: MIT Press.

George, E. W., Mair, H., & Reid, D. G. (2009). *Rural tourism development: Localism and cultural change*. Bristol, UK: Channel View Publications.

Globe. (2011). *An analysis of the economic development opportunities associated with the Green Economy in Newfoundland and Labrador*. http://www.exec.gov.nl.ca/exec/cceeet/publications/Green_Economy_NL.pdf. Accessed 2 July 2014.

Government of Newfoundland and Labrador. (2009). *Uncommon potential: A vision for Newfoundland and Labrador tourism*. http://www.tcr.gov.nl.ca/tcr/publications/2009/Vision_2020.pdf. Accessed 9 January 2012.

Government of Newfoundland and Labrador. (2012). *Oil and Gas*. http://www.economics.gov.nl.ca/E2012/OilAndGas.pdf. Accessed 2 July 2014.

Hadden, J. (2015). *Networks in contention*. New York: Cambridge University Press.

Hall, C. M., Müller, D. K., & Saarinen, J. (2009). *Nordic tourism: Issues and cases*. Bristol, UK: Channel View Publications.

Hall, P. A., & Soskice, D. (2001). An introduction to varieties of capitalism. In P. A. Hall & D. Soskice (Eds.), *Varieties of capitalism: The institutional foundations of comparative advantage* (pp. 1–70). Oxford, UK: Oxford University Press.

Hannigan, J. (2017). Toward a sociology of oceans. *Canadian Review of Sociology, 54*(1), 8–27.

Hansen, K. G. (2014). Greenlandic perspectives on offshore oil and gas activities: An illustration of changes in legitimacy related to democratic decision processes. *Journal of Rural and Community Development, 9*(1), 134–154.

Henley, J. (2019). *Centre-left Social Democrats victorious in Denmark elections*. https://www.theguardian.com/world/2019/jun/05/centre-left-social-democrats-set-to-win-in-denmark-elections. Accessed 19 November 2019.

Holzer, B. (2010). *Moralizing the corporation: Transnational activism and corporate accountability*. Cheltenham, UK: Edward Elgar Publishing.

Hughes, G. (2014). The energy sector in Scotland's future. *Oxford Review of Economic Policy, 30*(2), 374–391.

Huijbens, E. H. (2016). Topological encounters: Marketing landscapes for tourists. In G. T. Jóhannesson, R. van der Duim, & M. C. Ren (Eds.), *Tourism encounters and controversies: Ontological politics of tourism development* (pp. 201–218). London: Routledge.

Huijbens, E. H., & Jóhannesson, G. T. (2019). Tending to destinations: Conceptualising tourism's transformative capacities. *Tourist Studies, 19*(3), 279–294.

Jacobsen, J. K. S. (2015). North Cape: In the land of the midnight sun. In H. Nieves & S. R. Roseman (Eds.), *The tourism imaginary and pilgrimages to the edges of the world* (pp. 120–140). Bristol, UK: Channel View Publications.

Jennings, A. (2015). The old rock's rockin': A cultural anatomy of Shetland. In K. Vodden, R. Gibson, & G. Baldacchino (Eds.), *Place peripheral: Place-based development in rural, island, and remote regions* (pp. 156–179). St. John's, NL: ISER Books.

Kaltenborn, B. P., Linnell, J. D. C., Thomassen, J., & Lindhjem, H. (2017). Complacency or resilience? Perceptions of environmental and social change in Lofoten and Vesterålen in northern Norway. *Ocean and Coastal Management, 138*, 29–37.

Kinsella, S. (2018). *Husky'deeply sorry' for oil spill last month off Newfoundland*. https://www.cbc.ca/news/canada/newfoundland-labrador/husky-sorry-for-spill-changing-policy-1.4939450. Accessed 27 January 2020.

Kristoffersen, B. (2014). *Drilling oil into Arctic minds? State security, industry consensus and local contestation*. (PhD). Tromsø, Norway: University of Tromsø.

Kristoffersen, B., & Dale, B. (2014). Post petroleum security in Lofoten: How identity matters. *Arctic Review, 5*(2), 201–226.

Ladrech, R., & Little, C. (2019). Drivers of political parties' climate policy preferences: Lessons from Denmark and Ireland. *Environmental Politics, 28*(6), 1017–1038.

Larsen, E. R. (2006). Escaping the resource curse and the Dutch disease? When and why Norway caught up with and forged ahead of its neighbors. *American Journal of Economics and Sociology, 65*(3), 605–640.

Lijphart, A. (2012). *Patterns of democracy: Government forms and performance in thirty-six countries* (2nd ed.). New Haven, CT: Yale University Press.

Lindberg, F., Fitchett, J., & Martin, D. (2019). Investigating sustainable tourism heterogeneity: Competing orders of worth among stakeholders of a Nordic destination. *Journal of Sustainable Tourism, 27*(8), 1277–1294.

Locke, W. (2011). Newfoundland and Labrador, from austerity to prosperity–and back to austerity (?): Planning to avoid a financial crisis. *Newfoundland Quarterly, 104*(2), 39–41.

Longo, S. B., & Clark, B. (2016). An ocean of troubles: Advancing marine sociology. *Social Problems, 63*(4), 463–479.

Lund, K. A., Kjartansdóttir, K., & Loftsdóttir, K. (2018). "Puffin love": Performing and creating Arctic landscapes in Iceland through souvenirs. *Tourist Studies, 18*(2), 142–158.

Lund, K. A., Loftsdóttir, K., & Leonard, M. (2017). More than a stopover: Analysing the postcolonial image of Iceland as a gateway destination. *Tourist Studies, 17*(2), 144–163.

Milne, R. (2016). DONG energy's debut sparks outrage in Denmark over Goldman windfall. *The Financial Times*. https://www.ft.com/content/327d093e-2cd1-11e6-bf8d-26294ad519fc. Accessed 18 June 2018.

Milne, R. (2017). Oil and the battle for Norway's soul. *The Financial Times*. https://www.ft.com/content/c2dad93c-7192-11e7-aca6-c6bd07df1a3c. Accessed 17 April 2018.

Nilsson, J. H. (2019). Urban bicycle tourism: Path dependencies and innovation in Greater Copenhagen. *Journal of Sustainable Tourism, 27*(11), 1648–1662.

Norgaard, K. M. (2011). *Living in denial: Climate change, emotions, and everyday life*. Cambridge, MA: MIT Press.

Oklevik, O., Gössling, S., Hall, C. M., Jacobsen, J. K. S., Grøtte, I. P., & McCabe, S. (2019). Overtourism, optimisation, and destination performance indicators: A case study of activities in Fjord Norway. *Journal of Sustainable Tourism, 27*(12), 1804–1824.

Óladóttir, O. Þ. (2017). *Tourism in Iceland in figures*. Reykjavik: The Icelandic Tourist Board.

Olsen, L. S. (2016). Sami tourism in destination development: Conflict and collaboration. *Polar Geography, 39*(3), 179–195.

Ommer, R. E. (2007). *Coasts under stress: Restructuring and social-ecological health*. Montreal: McGill-Queen's University Press.

Ounanian, K. (2019). Not a 'museum town': Discussions of authenticity in coastal Denmark. *Journal of Tourism and Cultural Change, 17*(3), 285–305.

Overton, J. (1996). *Making a world of difference: Essays on tourism, culture and development in Newfoundland*. St. John's, NL: ISER Books.

Oxburgh, R. (2016). *Lowest cost decarbonisation for the UK: The critical role of CCS. Report to the Secretary of State for Business, Energy and Industrial Strategy from the Parliamentary Advisory Group on Carbon Capture and Storage (CCS)*. http://www.ccsassociation.org/news-and-events/reports-and-publications/parliamentary-advisory-group-on-ccs-report/. Accessed 4 June 2018.

Prince, S. (2017). Craft-art in the Danish countryside: Reconciling a lifestyle, livelihood and artistic career through rural tourism. *Journal of Tourism and Cultural Change, 15*(4), 339–358.

Quartz + Co. (2012). *The Danish oil and gas sector's development and social impact (1992–2022)*. Copenhagen: Quartz & Co.

Ramos, H. (2008). Opportunity for whom? Political opportunity and critical events in Canadian aboriginal mobilization, 1951–2000. *Social Forces, 87*(2), 795–823.

Ren, C. B., Bjørst, L. R., & Dredge, D. (2016). Composing Greenlandic tourism futures: An integrated political ecology and actor-network theory aapproach. In M. Mostafanezhad, R. Norum, & E. J. Shelton, & A.

Thompson-Carr (Eds.), *Political ecology of tourism: Community, power and the environment* (pp. 284–301). London: Routledge.

Roberts, T. (2019). *Near-record investment expected as 'bow wave' of activity heads for Newfoundland offshore*. https://www.cbc.ca/news/canada/newfoundland-labrador/nalcor-noia-exploration-investments-1.5316369. Accessed 27 January 2020.

Scottish Government. (2017). *Oil and gas production statistics 2016–17*. Edinburgh: The Scottish Government.

Scottish Governmnet. (2018). *First Minister's speech at Scottish Tourism Alliance*. https://www.gov.scot/publications/first-ministers-speech-at-scottish-tourism-alliance/. Accessed 28 January 2020.

Shadian, J. M. (2014). *The politics of Arctic sovereignty: Oil, ice, and Inuit governance*. London: Routledge.

Sinclair, P. R. (2011). *Energy in Canada*. Don Mills, ON: Oxford University Press.

Smith, J. R. (2016). *When petro-capitalism comes knocking: Community interpretations and responses to the Gros Morne fracking controversy* (MA). St. John's, NL: Memorial University of Newfoundland.

Sodero, S., & Stoddart, M. C. J. (2015). A typology of diversion: Legitimating discourses of tourism attraction, oil extraction and climate action in Newfoundland and Labrador. *Environmental Sociology, 1*(1), 59–68.

Søholt, S., Stenbacka, S., & Nørgaard, H. (2018). Conditioned receptiveness: Nordic rural elite perceptions of immigrant contributions to local resilience. *Journal of Rural Studies, 64*, 220–229.

Sovacool, B. K., Burke, M., Baker, L., Kotikalapudi, C. K., & Wlokas, H. (2017). New frontiers and conceptual frameworks for energy justice. *Energy Policy, 105*, 677–691.

Statistics Denmark. (2017). *Overnight stay by type of overnight accommodations, region, nationality of the guest, and period*. http://www.statbank.dk/TURIST. Accessed 11 November 2017.

Statistics Iceland. (2019). *Tourism short-term indicators in July*. https://www.statice.is/publications/news-archive/tourism/tourism-short-term-indicators-in-july/. Accessed 22 January 2020.

Stoddart, M. C. J., & Graham, P. (2016). Nature, history, and culture as tourism attractors: The double translation of insider and outsider media. *Nature and Culture, 11*(1), 22–43.

Stoddart, M. C. J., & Sodero, S. (2015). From fisheries decline to tourism destination: Mass media, tourism mobility, and the Newfoundland coastal environment. *Mobilities, 10*(3), 445–465.

Thorpe, H. (2012). Transnational mobilties in snowboarding culture: Travel, tourism and lifestyle migration. *Mobilities, 7*(2), 317–345.

Uldam, J., & Askanius, T. (2013). Calling for confrontational action in online social media: Video activism as auto-communication. In B. Cammaerts, A. Mattoni, & P. McCurdy (Eds.), *Mediation and protest movements* (pp. 159–178). Bristol, UK: Intellect.

Urry, J. (2013). *Societies beyond oil: Oil dregs and social futures.* London: Zed Books.

Urry, J., & Larsen, J. (2011). *The tourist gaze 3.0.* Los Angeles: Sage.

Vasi, I. B. (2011). *Winds of change: The environmental movement and the global development of the wind energy industry.* New York: Oxford University Press.

VisitDenmark. (2016). *Tourism in Denmark.* http://www.visitdenmark.dk/sites/default/files/vdk_images/PDF-and-other-files/Analyser/2016/turismen_i_danmark_-_engelsk.pdf. Accessed 11 November 2017.

VisitNorway. (2019). *The joy of different seasons.* https://www.visitnorway.com/plan-your-trip/seasons-climate/. Accessed 28 January 2020.

VisitScotland. (2014). *An overview of our target segments.* Edinburgh: VisitScotland.

VisitScotland. (2016). *Tourism employment in Scotland.* Edinburgh: VisitScotland.

Wejs, A., Harvold, K., Larsen, S. V., & Saglie, I. L. (2014). Legitimacy building in weak institutional settings: Climate change adaptation at local level in Denmark and Norway. *Environmental Politics, 23*(3), 490–508.

Wettergren, Å., & Soneryd, L. (2017). Denmark: From a green economy toward a new eco-radicalism? In C. Cassegard, L. Soneryd, H. Thorn, & A. Wettergren (Eds.), *Climate action in a globalizing world: Comprative perspectives on environmental movements in the global north* (pp. 171–192). New York: Routledge.

Widener, P. (2018). Coastal people dispute offshore oil exploration: Toward a study of embedded seascapes, submersible knowledge, sacrifice, and marine justice. *Environmental Sociology, 4*(4), 405–418.

World Tourism Organization. (2017). *UNWTO Tourism Highlights* (2017th ed.). Madrid: World Tourism Organization.

3

Cultural Dimensions of the Oil-Tourism Interface

This chapter considers cultural variation in the oil-tourism interface. Cultural visibility is an important aspect of social power, which can be leveraged by oil sector, tourism sector, or social movement players to gain political efficacy. Cultural visibility can also help reinforce or potentially challenge the economic power of oil sector or tourism sector interests. Here, we focus on how culture is used to evaluate extractive and attractive development models, and how those evaluations give rise to specific configurations of social and symbolic boundaries—the oil-tourism interface—that vary across each of our five cases.

The oil-tourism interface is grounded in a relational approach to comparative cultural analysis (Emirbayer 1997; Lamont and Thévenot 2000; Lamont et al. 2016) that emphasizes how players use and deploy culture in institutionally constrained arenas to solve problems (Burkhus and Ignatow 2019; Leschziner 2019; Swidler 1986, 2013; Vaisey 2009), draw consequential distinctions between what is "good" or "bad" (Lamont 1992, 2012), and provide socially acceptable justifications and explanations (Boltanski and Thévenot 1999; Tilly 2006; Vaisey 2009). In "using" cultural tools—concepts, narratives, stories, rituals, ideologies, etc.—to these ends, people construct social and symbolic boundaries that

M. C. J. Stoddart et al., *Industrial Development and Eco-Tourisms*,
https://doi.org/10.1007/978-3-030-55944-1_3

shape a wide variety of important outcomes (Lamont and Molnár 2002; Lamont and Thévenot 2000). These cultural dynamics shape approaches to environmental policy and governance, as well as the construction of cultural blind spots or other forms of collective unawareness and socially organized denial (Friedman 2019; Norgaard 2011).

From this broader literature on repertoires of evaluation and boundary processes, we place a particular emphasis on how evaluations of tourism and oil development are differentially embedded within institutionalized "orders of worth" that appeal to broadly shared conceptions of the common good, and which mark certain types of things and actions (e.g. nature, environmental conservation; economy, increasing profits) as more or less valuable (Boltanski and Thévenot 2006; Thévenot et al. 2000; Thornton and Ocasio 2008). Boltanski and Thévenot (2006) identify 6 orders of worth, including *inspired worth*, which ascribes value to spiritual commitment, natural beauty, and indifference to material goods and money; *domestic worth*, which ascribes value to tradition, intimate relationships, and social hierarchies; *opinion worth*, which ascribes value to fame and popularity; *civic worth*, which ascribes value to collectives and shared conventions and culture; *market worth*, which ascribes value to wealth and profit; and *industrial worth*, which ascribes value to technical skill and efficiency, professionalism. To this initial list of 6, others have added an *ecological/green* order of worth that ascribes value to nature and promotes environmental protection and conservation and responses to climate change (e.g. Kukkonen et al. 2020; Lafaye and Thévenot 1993; Scott 2020; Ylä-Anttila and Kukkonen 2014; Ylä-Anttila and Luhtakallio 2016).

As might be expected, our cross-case comparisons reveal that general discussions of the environment tend to be characterized by inspirational worth (focused on the beauty of nature), but ecological and industrial orders of worth play an unambiguous role in shaping how tourism and oil development models are evaluated. Across cases, we see that oil development is primarily evaluated in terms of an industrial order of worth which emphasizes the role of technical innovations and the competence of sophisticated risk management. When it comes to evaluations of tourism development, we more often see invocations of an ecological order of worth that strongly emphasizes the importance of nature, understanding and protecting local ecologies, and responding to global climate change. In Chapter 1, we noted three facets of social-ecological relationships with oceans as travel pathways, natural resource pools, and objects of political contestation and scientific concern over ecological problems. The comparative cultural analysis in this chapter demonstrates how oceans are differently valued across our case study regions.

The *tourism + ecological worth* and *oil + industrial worth* pairings are broadly consistent across our cases, but as shown in Table 3.1, the relative importance of each varies.

The core findings reported in this chapter stem from qualitative and comparative analysis, triangulated with computational text analysis. We draw on the full range of data informing this book—interviews, ethnographic field notes, documents, websites, and news stories—to deepen our understanding of the cultural dimensions of the oil-tourism interface. Our multi-method approach is detailed in the Epilogue on Methodology, but we briefly introduce the basics of our computational text analysis to provide context for the results presented below.

Our computational text analysis uses a special class of topic models to analyse news stories on oil and/or tourism from each of the five cases. Topic models are widely used in cultural sociology, in part because of their affinities with relational theories of meaning and language (Blei 2012; DiMaggio et al. 2013; Light and Cunningham 2016; McLevey 2021; Mohr et al. 2013; Nelson 2020; Roose et al. 2018). Topic models are a class of machine learning models that enable us to identify and explore latent "topics" in unstructured text data. They assume that texts—in our case, news stories—are composed of multiple "topics" in different proportions. For example, a news story about the impacts of tourism development in rural settings might include "topics" for environment, wildlife, camping, employment, and economic benefits. Each of these topics consists of words that have different probabilities of occurring. The word "puffin," for example, might have a high probability of appearing in topics such as wildlife, nature, and the environment, but a low probability of appearing in a topic concerning employment.

In most cases, topic models are fully unsupervised, meaning that researchers inductively identify topics in a collection of texts to better understand their relationships to one another without specifying in

Table 3.1 Summary comparison of case study regions

	Oil development	Tourism development	Industrial worth	Ecological worth	Market worth	Inspirational worth	Key Blended Orders
Scotland	High	High	High	High	High	High	Industrial + Ecological
Norway	High	Low	High	Low	High	High	Industrial + Market
Newfoundland and Labrador	High	Low	High	Low	High	High	Industrial + Market
Denmark	Low	High	Low	High	High	High	Ecological + Market
Iceland	Low	High	Low	High	High	High	Ecological + Market

advance what one expects to find. Given that we seek to understand and compare *specific* discourses across cases in a way that blends inductive and deductive inquiry, we employed semi-supervised hierarchical models (Gallagher et al. 2017) that look for coherent topics around sets of focal concepts derived from our qualitative analysis. It accomplishes this by tracing cultural variation on both the meso-level—in the form of topics, comprised of clusters of cultural units (such as specific concepts)—and on the macro level, in the form of more general "discourses" that, when combined with our qualitative and comparative analysis, can help shed light on the broader orders of worth at play in each of our cases.

Below, we provide a more detailed discussion of the specifics of each case. Each begins with an overview of the key findings from our qualitative and comparative analyses and is then triangulated with results of a computational text analysis. The chapter concludes with a high-level comparison of cultural similarities and differences in the oil-tourism interface, and the relative importance of ecological and industrial orders of worth in each.

Complementary Orders of Worth in Scotland

When discussing the environment, our Scottish participants most frequently invoked the beauty of coastal landscapes; the iconic inland Highlands; wildlife such as seabirds, seals, whales, and dolphins; hiking; and beach walking. All of these illustrate an inspirational order of worth when discussing the environment in general. A participant from the tourism sector described the Scottish tourism image as follows: "If I had to generalize, it would be the perception of the Highlands, its mountains, its open moorlands. It's… you know, almost a medieval vision of, you know, like endless tracts of forests and stags" (SCO_TOU_01). A government participant similarly notes that "Wildlife tourism [seabirds, seals] is recognized as being really important for Scotland" (SCO_GOV_07). This way of conceptualizing the environment is more clearly aligned with nature-oriented tourism than with oil and other extractive industry and tends towards an inspirational order of worth that stresses iconic nature and wildlife.

For many of our interview participants, the positive impacts of tourism were primarily framed in terms of economic benefits, including as providing ways to make a living in rural communities, and by extension emphasize market worth. As a tourism sector participant puts it: "So, tourism can be a catalyst for social sustainability in terms of creation of jobs, creation of enterprise growth, which allows communities to exist in areas where, maybe inward investment is more difficult for other areas" (SCO_TOU_01). However, we also see a strong emphasis on ecological worth in the emphasis on environmental sustainability, especially when tourism development is compared to oil and other extractive industries. As the same participant puts it, "There is much more cognizance today … that there needs to be this balance between maintaining economic growth but to ensure that economic growth through tourism, you have to protect the environment" (SCO_TOU_01). We see a further indication of the role of ecological worth from participants who promote tourism as a tool for legitimating nature preservation initiatives and as opening opportunities for environmental education.

Our participants were also reflexive about the negative environmental impacts of tourism. They highlighted overcrowding of camping grounds, trail degradation, garbage and waste, the challenges of accommodating RVs, potential challenges to wilderness values, and the development of otherwise undesirable infrastructure. A government participant describes local environmental impacts as follows:

> I think perhaps the road infrastructure in a lot of places is not quite able to cope with the number of people. Obviously, if you keep building more roads that increases the impact on that area, so it's a bit of a difficult one to solve. Perhaps it can be solved by better public transport. … The National Nature Reserves, they have to manage visitors quite heavily or quite carefully because they get lots of soil erosion in certain areas and litter is an issue as well. (SCO_GOV_14)

The need to properly manage AirBnB and similar short-term rentals also came up as a potentially negative social impact. To a lesser extent, there are concerns about the impacts of tourism development on employment and labour in the context of Brexit, as well as the need to adapt to generational shifts as baby boomers age out of the tourism market and millennials become increasingly central (fieldnotes, March 17, 2017). In general, we see a shared understanding that the localized environmental and social impacts can be managed with more effective tourism governance and codes of conduct. These understandings reflect a blend of ecological, market, and civic orders of worth in evaluating tourism development.

Participants also emphasized more global impacts, the most salient of which is the contribution of tourism to climate change due to the carbon intensity of travel by cars, buses, planes, and cruise ships. For example, a government participant notes:

> I suppose more and more people are coming in by plane, which environmentally—there's issues there around emissions. I suppose that's true of every [place] which has a tourism sector. Beyond that, it's difficult to see, I think. Environmental regulations are being tightened all the time, I think more and more people are conscious about these things. (SCO_GOV_05)

This quote captures the tensions inherent to tourism between locally sustainable development and the broader issues of carbon-intensive transportation.

Despite the strong emphasis on tourism in talk about the environment, the Scottish case is characterized by an emphasis on economics and market worth, including the place of oil development in the potential economic situation of an independent Scotland. When it comes to oil and climate change, the main themes centre on government policy responses, risks and impacts, and the notion of the "energy trilemma," which we discuss below. This shifts the focus from a discussion of ecological worth to industrial worth by emphasizing the importance of technologically innovative responses to climate change.

Much of the discourse around the Scotland's oil sector is very positive, explicitly linked with economic development, positioned as an important part of regional history and culture, viewed as a cause of migration to Scotland, and connected to the potential for Scottish independence. For example, there is an emphasis on how the oil sector transformed

Aberdeen from a fishing centre to a more prosperous oil city following the decline of the fishery. An oil sector participant sums up the positive impacts as follows:

> Before oil Aberdeen was quite a rural, probably quite homogenous, probably quite—you know, parochial, maybe? As I say, I'm not a sociologist. But now it's a much more diverse city and region. Lots of different cultures, lots of different experiences. (SCO_OIL_19)

One particularly visible example of the centrality of oil in talk about the environment is at the Aberdeen Maritime Museum, which features exhibits that portray coastal environments as historically and culturally important sites of oil extraction, and as a focus of corporate environmental responsibility. Indeed, part of the narrative is that the Scottish North Sea is largely a post-peak seascape, so issues of decommissioning, environmental sustainability, and renewable energy transitions are increasingly salient (fieldnotes, March 6, 2017).

Our participants also describe the drawbacks of oil development, including negative economic impacts associated with the volatile nature of oil prices, and the oil price declines of recent years. As a government participant puts it, "It's brought a lot of wealth and prosperity. When [...] the price falls, it brings a lot of hardship and heartache. ... In the last couple of years, people who have found it very difficult to live with redundancy. People who have held very highly paid jobs. And suddenly got nothing to do" (SCO_GOV_06). Another government participant makes similar observations about Aberdeen as a host community:

> But certainly, Aberdeen when the oil price is higher—Aberdeen is prosperous. House prices are high, lots of new shops open, all those sorts of luxury businesses that people can afford when things are going well. They all do very well, restaurants do very well. As soon as it drops everything starts to close, that's what's happening in the last few years, and you can see that it's very closely — the value, the welfare — is very closely related to the oil price. (SCO_GOV_14)

The price of oil has since dropped even more dramatically in the context of the COVID-19 global pandemic, which began after our data collection was completed.[1] Our participants also emphasized environmental risks (especially for seabirds), biodiversity issues, controversy over fracking, and contributions to climate change, but these issues were less central than issues that were more clearly oriented towards economics and market worth.

Despite acknowledging these and other risks, the general perception is that risk mitigation is well done and technological innovations promise positive responses to climate change, further illustrating the general process of blending industrial and ecological worth. The famous Brent Spar conflict is invoked—for example in the Aberdeen Maritime Museum—as a moment where the Scottish oil sector began taking environmental responsibility more seriously. In fact, participants from the Scottish oil industry described how oil development could be done in ways that respond positively to climate change, such as by lowering the carbon intensity of oil extraction through technological innovations. For example, an oil sector participant states:

> I think that people do recognize, and the industry, I think, itself broadly recognizes, the need to move to lower carbon. And of course, some of the technologies and hardware can be used for renewables as well. So, there's know-how that can be utilized for offshore wind, etc. (SCO_OIL_19)

More generally, oil sector data—especially from Shell, Oil & Gas UK, and BP—makes extensive use of environmental sustainability discourse that emphasizes decommissioning old wells and mitigating oil risks, while also asserting the potential coexistence of fossil fuels and renewables. For others, such as web content from Equinor, responses to climate change are linked to talk of transitions away from oil and towards "energy" more broadly. In this way, ecological and industrial worth are further blended.

The "energy trilemma" is a challenge connecting many of these themes in the Scottish case: (1) the world has a growing demand for energy, (2) oil is a major historical driver of economic development and wellbeing

---

[1]The timing of our data collection is relevant here. Setting aside the impact of the COVID-19 global pandemic on oil prices, the interviews for the Scottish case took place when oil prices were already in decline. This likely influenced the emphasis on volatility.

in Scotland, and (3) we are living in an increasingly carbon-constrained world that demands aggressive climate action. The challenge is how to balance these three competing priorities. Decommissioning is an important part of this struggle, including how to do so in an environmentally responsible manner, and whether the industry or the public sector should bear the financial burden.

We triangulate our findings from the qualitative analysis of the interviews, field notes, documents, and websites with topic models of news stories. In the Scottish case, we find that independence and Scotland's relationship with Great Britain is the dominant theme in this data. The terms most strongly associated with this topic are unambiguous: independence, referendum, independent, westminster, vote, union, alex_salmond, political, voter, snp. The second most dominant topic focuses on the rise and fall of oil prices, reflecting concerns with volatility raised by our interview participants and illustrative of the importance of the market order of worth. Following these two topics, we find many distinct topics on a range of issues, with most focusing on energy (including renewables), tourism (including hospitality and differences between rural and urban tourism), and finally nature and the environment.

When we look at the results of the hierarchical topic model, shown in Fig. 3.1, we see that each specific topic is clustered into more general discourses that maintain a clear separation between tourism development and oil development. This is the case in nearly all instances, which supports the general finding that these two development models are treated as parallel pathways in Scotland.

The central unlabelled node in Fig. 3.1 is the "top" of the hierarchy connecting all discourse clusters. The width of each black line indicates the strength of the association between topics and clusters. The black nodes with white numbers identify the discourse clusters, within which each topic is identified by a number and a list of the most important words and bigrams.

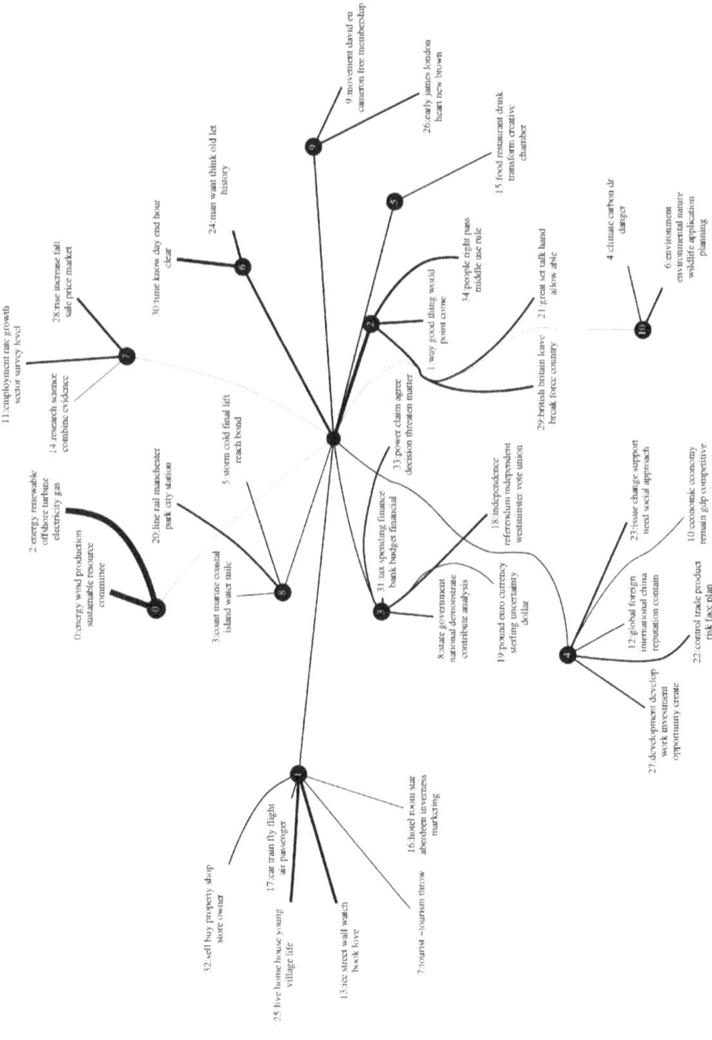

Fig. 3.1 Macro-level discourse network for Scotland

While this topic model does include some "junk topics"—which, because these models are semi-supervised, is a finding[2]—the key takeaway for our purposes is that none of the discourse clusters bring talk about tourism development and oil development together, and no concepts or terms that are associated with either type of development model appear together in any other topics, such as economics and climate change.

In sum, across all our data sources we see persistent patterns. Talk about the environment tends to emphasize inspirational worth, focused as it is on iconic seascapes and wildlife. Nature-oriented tourism dominates talk of the environment specifically, and evaluations of tourism development are clearly embedded in a more general ecological order of worth. However, discussions of oil are more central to the Scottish case, which is partly due to the historic role that oil played in the Scottish economy, the current "energy trilemma," and the social and economic benefits of oil development given the potential future of an independent Scotland. For the most part, the discourse around oil in Scotland is positive, with an emphasis on technological innovation, environmental sustainability, and the gradual shift to renewable energies, during which many believe fossil fuels and renewables will coexist. In this case, industrial worth is clearly dominant, with a secondary focus on market worth. We find few explicit connections between tourism and oil development models in the Scottish case, though industrial and market worth are occasionally invoked in evaluations of tourism, and ecological worth is sometimes invoked in assessments of oil development. Overall, oil and tourism are conceptualized as complementary and parallel development pathways. Ecological and industrial worth appear in complementary ways and are occasionally blended.

[2]Our semi-supervised learning approach gently nudges the topic model towards a list of focal concepts that are shared across the models for all five cases. If any of those focal concepts returns a nonsensical "junk" topic, it means that there is no coherent topic or theme that forms around that focal concept in the specific case. Topic 24 (with keywords on the upper left of Fig. 3.1 is such an example).

The Dominance of Industrial Worth in Norway and Newfoundland and Labrador

Norway

The sea and coastal environments are central components of Norwegian culture, spanning from Viking history and folklore, to rural fisheries, to iconic fjords and seascapes. Despite the obvious presence of a nature-oriented inspirational order of worth when discussing the environment, Norway differs substantially from our other cases in that talk of coastal environments is dominated by considerations of oil extraction and oil prospectivity, particularly in the Norwegian Sea and increasingly in the Barents Sea. In general, we found that the oil industry is valued and seen as economically significant, technologically innovative, and an important part of Norwegian history and culture. As a government participant put it:

> This [the oil sector] is of immense importance for Norway being by far Norway's largest and by far the most profitable industry. So, it creates jobs all along the coast. And it generates income to the national treasury that finances education, transportation, healthcare. Large part of Norway. So, it's of immense importance for national income. (NWY_GOV_02)

This positive assessment of the economic and social impacts of oil development is shared by many environmental participants, despite their critical views of future oil development. As an ENGO participant puts it, "Positive impact would be that we have a lot of money. We're the richest country on the planet, as they say, because we have the oil fund.[3] Although that's equally as much about sound political decisions as it is about oil. And, you know, you have a lot of jobs" (NWY_NGO_02). Oil development is seen as having positive social and economic impacts on local communities and is seen as a source of pride, especially when it comes to Statoil as a public oil company (now rebranded as Equinor,

[3] Discussed in Chapter 2.

which is a hybrid public-private company). The centrality of the market order or worth is clearly on display here.

As we expect given the role of oil in Norway's political economy, many of our participants saw the Norwegian oil sector as a major economic driver and as the foundation for the country's social welfare system and high quality of life. While some have concerns about dependence on oil and contributions to climate change, Norway is viewed by many as a global leader in terms of technological innovation, as well as being more socially responsible and environmentally sustainable than elsewhere in the world. Like Scotland, then, we see appeals to industrial worth in response to environmental concerns that are rooted in ecological worth. For example, a government participant notes:

> We have a very long tradition in Norway for really driving [health and environmental] standards in the industry for decades, really. When it comes to emissions to sea there was strict regulations ... for emissions to sea. We've already [had], since the early 1990s, a very high CO_2 tax for emissions to air, climate greenhouse gas emissions. (NWY_GOV_02)

Further illustrating this uneven blending of orders of worth, many of our participants emphasize the potential of technological solutions to climate change, including transitions to renewable energies, carbon capture and storage, efforts to displace "dirtier" fossil fuels from the global energy system, and running biodiversity protection programmes. This re-envisions the oil sector more broadly as an "energy sector" in which oil and renewable energies might coexist.

Despite this positivity and technological optimism, our participants also articulated environmental risks associated with oil, and invoked ecological worth. For the most part, the central issues are contributions to climate change or the risks to fisheries and vulnerable Arctic ecologies. An ENGO participant describes their opposition to the expansion of oil development as follows:

> The fisheries impact. But also, the tourism. You have a lot of tourism in that area [Lofoten] as well, which will suffer if you go for it. ... because the fisheries are our second biggest income in Norway the one that you

always can compare to oil and gas industry. And, of course, you have the world's biggest cod stock in Lofoten. … And you have the biggest coldwater and deepwater coral reefs in that area. You have the biggest sea bird colony on mainland Europe in that area. So, there's vast numbers of important biological species there and habitats. (NWY_NGO_04)

When it comes to tourism development, on the other hand, "being close to nature" is frequently invoked as a tourism attractor but also as characteristic of Norway and Norwegianness. When asked about what draws tourists to the region, an ENGO participant replies, invoking inspirational worth: "It's, of course, the nature. And the natural experiences. The Lofoten Islands. Its majestic. And looks really beautiful. I think that's the one thing that draws people the most" (NWY_NGO_02).

Our participants focus on the beauty of coastal environments, fjords, mountains, the Northern Lights, and wildlife (especially whales and seabirds), and describe interacting with the environment by hiking, boat tours, and cruises. This inspirational worth is blended with market worth, with tourism seen as yielding positive economic benefits to host communities, including remote communities like Lofoten. For example, a government participant emphasizes the importance of tourism development for rural communities by saying, "Many of the municipalities and small societies have challenges about having enough jobs. They're getting fewer. We don't have enough interesting jobs for the young. They're moving away from the islands and that. I would say that the tourist industry can be an answer" (NWY_GOV_16). In addition, the potential of tourism-fisheries synergies is often emphasized, as is the potential for using tourism as a tool for environmental education.

Our participants were cognizant of tourism's limitations and negative consequences in the form of overcrowding of tourism spaces, travel routes, localized waste management issues, or the displacement of rental housing by an ascendant AirBnB. For example, a government participant tells us:

We have people coming to Lofoten all year around now—there won't be time for the nature to restore itself. … So, I think we'll see the scars of this summer for a long time because people will just continue to come as

> well. So, if we don't do anything it won't be sustainable anymore, I think. (NWY_GOV_18)

Another government participant similarly emphasizes the infrastructure gap in dealing with the negative local impacts of tourism: "The hikes through the mountains don't have the infrastructure, don't have sanitary facilities. You cannot get rid of the garbage. They're overcrowded with—the municipalities don't have a system where they empty the garbage containers" (NWY_GOV_16). As in Scotland, participants expressed the perceived need for better tourism governance, not a retreat from tourism development. In other words, our participants did not see tourism as inherently more sustainable than other development models; rather it must be managed properly. A government participant sums this up as follows, "We want the tourists. We want them to come here. We want them to experience the same things as we do every day. But the main problem is we don't have the infrastructure to cope with that number that is coming" (NWY_GOV_19). The carbon intensity of cruise ship tourism impacts on Arctic nature (e.g. Svalbard), and impacts on other natural areas came up as issues that need to be addressed. For the most part, these tourism governance concerns were raised by participants from the government and come up repeatedly in our field notes from research in Lofoten.

Once again, these findings are confirmed by our computational text analysis of news stories from Norway. In this case, our topic model reveals that the dominant topic is economic (the most associated words being: economy economic price financial sector growth bank currency market spending). This was followed by: government (the most associated words being government national election decision future policy risk party campaign source), climate change and the Arctic (the most associated words being arctic ice melt polar sea_ice greenland north_pole russia alaska canada), and a variety of coherent topics related to offshore drilling, renewable energies, climate change and climate science, international markets, tourism, restaurants, hotels and hospitality, travel, and extreme weather.

Our hierarchical topic model analysis, the results of which are shown in Fig. 3.2, again reflect the findings from the qualitative analysis. We

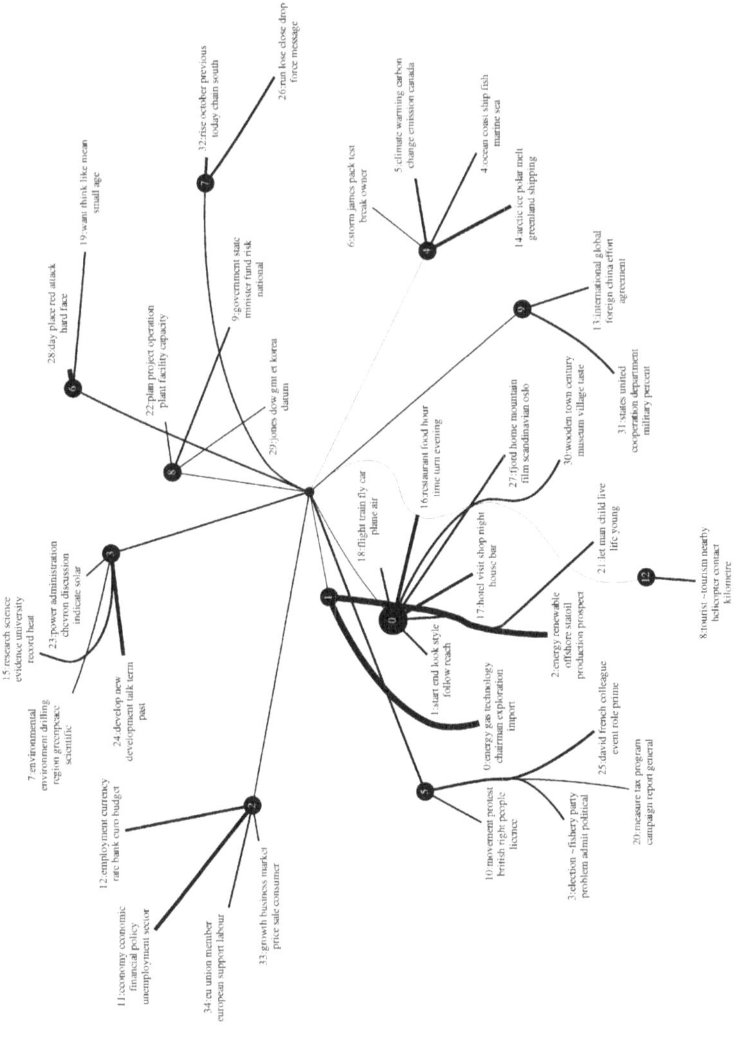

Fig. 3.2 Macro-level discourse network for Norway

see clusters related to science and technology (e.g. cluster 3); climate change, the Arctic, and marine ecologies (e.g. cluster 4); energy, oil, and gas (e.g. cluster 1); hotels, restaurants, hospitality, and mobilities (e.g. cluster 0). Among other things, this analysis shows that talk of oil in Norway is more tightly connected to talk of energy, renewables, scientific and technological innovation than in Scotland, and that it tends not to intersect with tourism-related topics. We see some nature-oriented topics (e.g. topic 27 in cluster 0) that get clustered with tourism topics, but none that are clearly clustered with oil or energy. Instead, we see a concern with environmental sustainability and renewable energies clustered with talk of oil. Otherwise, most of the topics focused on the environment are grouped into a discourse cluster on climate change and the Arctic.

Overall, oil is unambiguously the dominant development pathway in Norway and the industrial and market orders of worth dominate. However, there is also a strong focus on environmental sustainability, technological innovation, climate science, and the notion of transitioning from an "oil" industry to an "energy" industry where "cleaner" oil and renewables coexist. As in Scotland, this represents a blended industrial-ecological order of worth in which technological innovations (industrial worth) are offered as solutions to problems framed in terms of the ecological order of worth.

Tourism is clearly secondary to oil in Norway, as a mode of development, but is viewed in generally positive terms. However, we do see more extensive criticism of tourism development than in other cases, in part because it is also carbon-intensive and poses additional environmental risks. For these and other reasons, there is a focus on the importance of managing tourism properly. The Norwegian Arctic is one of the key places where these two ways of conceptualizing coastal environments as sites of development intersect. This reveals tensions related to the impacts of climate change and the role that tourism can play in environmental education.

Newfoundland & Labrador

The second case that is dominated by oil development + industrial worth is Newfoundland and Labrador. However, as we see in other cases, the most common ways of conceptualizing coastal environments and communities unambiguously invoke inspirational worth; they focus on seascapes, coastlines, whales, puffins, seabirds, icebergs, Gros Morne national park, Fogo Island Inn, food, drink, the authenticity of rural communities, and local arts, crafts, and culture. The concept of "being close to nature" is frequently invoked by participants who define nature-oriented tourism in the province as peaceful and "off the beaten track." For example, when asked about what draws visitors to the region, an ENGO participant notes:

> Probably the beauty of the landscape… Gros Morne obviously is a huge hub in this particular area. And then a lot of the other tourism operators have sort have tagged onto the image that Gros Morne is portraying on the west coast. And when you look at the messages that are being sent out by the government, Newfoundland and Labrador, as far away from Disneyland as you ever want to be, I think that sort of hits the nail on the head. People are trying to get away. (NL_NGO_16)

This tourism-centred depiction of the coastal environment is closely connected to talk of hiking, boat tours and whale watching, and environmental education. Environmental sustainability concepts are also frequently invoked, especially as they relate to the negative environmental impacts of snowmobiling, off-highway vehicles, or fracking. Together, then, discussions of nature-oriented tourism are characterized by an orientation towards inspiration and ecological orders of worth. However, participants across all sectors also invoked market and civic orders of worth by emphasizing the positive economic impacts of tourism and social benefits for rural communities. For example, a government participant notes, "I think among some of the benefits are that tourism can not only survive but maybe even thrive in some of our rural settings and for the most part western Newfoundland is a rural setting. It's got tremendous appeal as is. We don't need to be something we're not. It's a unique place and a special place in the world" (NL_GOV_07). Similarly, a tourism sector participant says:

> I don't see very many drawbacks; I do see a lot of advantages. You know? As an industry, it can be long term. It's renewable. It's pretty well non-damaging environmentally. Certainly, those are all advantages that [you don't] see with oil, you know? The product is long term. It's low impact. It's actually supporting nature and species in their habitat, you know, whales, seabirds. And raising community awareness, world awareness of species of wildlife. (NL_TOU_12)

As illustrated by this interview excerpt, environmental sustainability and education was also a dominant theme in our interview data. This comes up in reference to places like Gros Morne national park, Petty Harbour mini-aquarium, and the Johnson Geocentre. The emphasis tends to be on coastal ecologies and geology. There is also a recurrent emphasis on sustainability, though this tends to focus on local environmental impacts more than larger problems such as global climate change. As we have seen in some other cases, our participants shared a general sense that tourism affords opportunities to legitimate nature preservation initiatives and to benefit communities beyond increased revenue and employment, for example by supporting local art and crafts.

There are fewer critiques of tourism development than we see in some other cases, such as Norway and Iceland. However, our participants raise concerns about the threat of overtourism, the integrity of local environments, and the importance of building appropriate infrastructure. For example, a tourism participant points to the localized environmental impacts as follows: "Particularly in this region when you have that much tourism coming within a span of three to four months and growing from the first of mid-May to beginning of June until September, it has an impact on the infrastructure of the local community, which is really small. Garbage, water that kind of stuff" (NL_TOU_18). An ENGO participant similarly points to the localized environmental impacts, as well as the challenges of seasonality and labour to doing sustainable tourism development as follows:

> And of course, a downside, overuse is an issue. Too much use of trails, too much driving, too many campers, can actually diminish whatever conservation we're trying to achieve in these regions. I guess the seasonal nature in northern climates is an issue. So, what do you do the next rest of the year if you're reliant on working in a restaurant or working in the park in the summer, what do you do the rest of the time? (NL_NGO_03)

Turning to oil development, the dominant theme is positive economic benefits and employment opportunities, illustrating a focus on market worth. Indeed, oil has been a major economic driver for the province over the past 20 years, allowing the recovery from the collapse of the cod fishery and the transition from a "have not" to a "have" province within Canada.

Once again, we see concerns rooted in notions of ecological worth being addressed by shifts to solutions that have industrial worth. For example, oil is commonly seen as a source of environmental risk, but usually as a prelude to, or in conjunction with, framing oil risk mitigation and governance in the province as sustainable and well-done (the kind of engineering and technical competence that defines the industrial order of worth). For example, when asked about oil and sustainability, an oil sector participant replies:

> I think that's a sustainable industry in that way and you can never have an industry like the oil industry without having some impact on the environment. But it's how you manage it that's the important thing. I think a great deal of effort goes into [it] and oil operators are very conscious of the negative impacts if they don't do their jobs responsibly. (NL_OIL_04)

The dominant narrative is of oil as positive, economically significant, and sustainable.

The main critical counter-discourse we see is that oil development poses significant environmental risks for whales and seabirds and is in direct conflict with the ecological values promoted in touristic and educational spaces like Gros Morne National Park. This critique is anchored in notions of ecological worth. Despite a generally positive view of the oil sector, a government participant notes that there need to be boundaries around appropriate oil exploration:

> The exploratory areas that need to come off the list first are all of those that pose a significant risk to the environment and other relatively more sustainable economies. So, the Gulf of St. Lawrence should be right off the list. It's a semi-enclosed sea, it's got counterclockwise circulation, it's got ice covering the winter … It touches five provinces. Oil platforms are a continuous source of small spills and they pose an unacceptable threat. … And do you think tourists will really want to come and enjoy that? (laughs) I don't think so! (NL_GOV_31)

As this excerpt illustrates, there are conflicts over oil development impinging on valued and established tourism and fisheries economies, with Gros Morne and the Gulf of St. Lawrence being the most contested spaces. An ENGO participant sets up this juxtaposition by saying, "I feel oil is a short-term boom and bust industry that is endangering the survival of our sustainable, long-term, unique, coastal sustainable inshore fisheries and tourism industries" (NL_NGO_09). However, these criticisms of oil development tend to focus on specific projects rather than the oil sector overall. More generally, some participants—all external to the oil sector—shared concerns about fossil fuel contributions to climate change, fracking proposals for the west coast of Newfoundland, seismic impacts from exploration activity, and the risk of oil spills, which is often envisioned in reference to the 2010 Gulf of Mexico spill. In addition to these more environmental concerns, some participants raise concerns about price declines, volatility, and the need to consider restructuring the global industry.

As in our other cases, there is talk in Newfoundland and Labrador of renewable energy transitions. At oil sector events, for example, speakers from Statoil (now Equinor) and BP support renewable energy development as part of a suite of responses to climate change that also includes technological innovation, carbon capture and storage, and shaping consumer behaviour (fieldnotes, June 21, 2016; June 22, 2017). Government websites and documents also point to renewable energy development, sometimes mixing talk about mega-hydro (i.e. the Muskrat Falls project in Labrador) and other forms of renewables (i.e. wind) with the idea that oil revenues can support the development of renewables.

Wildlife and icebergs are, in a sense, boundary objects between these two development models. On the one hand, whales, seabirds (e.g. puffins), fish, and other wildlife are portrayed as key tourism attractors vulnerable to the risks posed by oil development, disasters, and exploratory seismic activity. Whereas icebergs appear as key tourism attractors in the tourism imaginary, in the oil imaginary they are defined as sources of risk to be navigated.

Many of these findings are reinforced in our analysis of the news story data. The two most dominant themes are economic: food, restaurants, and accommodations; and mobilities (cars, planes, boats, trains, etc.). In addition to these, we see several topics related to oil (mainly offshore drilling and fracking); climate change and the environment; government, politics, and elections; work and employment; the Arctic; and the fishery.

Figure 3.3 shows the results of the hierarchical topic model, which shows—in stark terms—the separation of the oil and tourism development discourses. Cluster 0 encompasses the primary tourism topics, including food and restaurants, history, camping, and the cod fishery. Not all of these topics are coherent or easily interpretable, however, which suggests that some of the focal concepts used to seed the topic model across all five cases apply less to the case of Newfoundland and Labrador than they do in other cases. Discourse cluster 1 contains the oil development discourses, including topics related to economics, corporate social responsibility and environmental sustainability, and politics and elections. The one oil topic that is not part of this discourse cluster (id 3) is focused on fracking, drilling, and offshore oil. Again, this discourse cluster contains some junk topics, suggesting the general lack of relevance for some focal concepts in the Newfoundland and Labrador case. It is important to note, however, that this cluster also includes a topic related to Gros Morne National Park, which is a focal point for controversies over oil development. Finally, we note that talk of climate change is confined to discourse cluster 2, which also covers the Arctic.

Overall, the dominant pattern we see here is that oil and tourism are conceptualized as parallel and often completely separate development pathways, both of which have positive economic benefits and both of which have played important roles in the Newfoundland and Labrador economy—not to mention culture and identity—following the

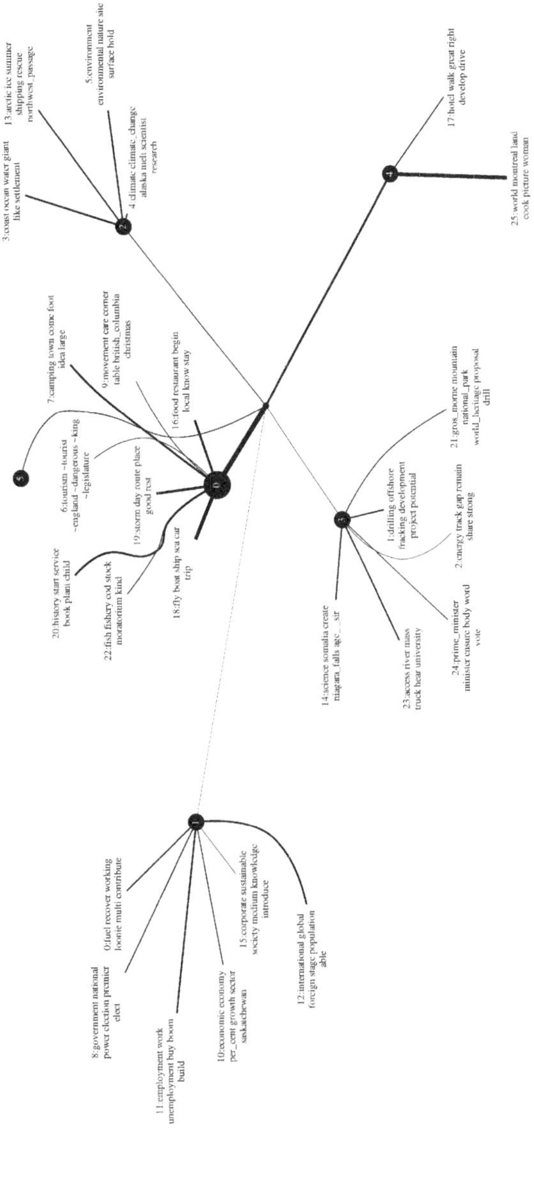

Fig. 3.3 Macro-level discourse network for Newfoundland and Labrador

collapse of the cod fishery. Beyond the inspirational worth of beautiful nature, we see a strong emphasis on industrial and market worth when discussing oil, and a strong focus on market worth and ecological worth when discussing tourism. Antagonism and the notion of tourism and oil as incompatible come up less often and is limited to specific episodes of contention where proposed oil development impinges on existing tourism landscapes and economies, particularly around Gros Morne in western Newfoundland.

The Dominance of Ecological Worth in Denmark and Iceland

Denmark

In the case of Denmark, talk of coastal environments and communities focuses predominantly on nature, coastlines and seascapes, wildlife such as shorebirds and seals, and an emphasis on cycling, hiking, swimming, and beach walking. All of this is anchored in inspirational worth. For example, when asked what draws tourists to the region, a tourism sector participant replies:

> Wide beaches. I have absolutely no doubt in my mind. … So that's why people come here. And if they come here, oh there's a nice museum, oh there's a nice town or whatever, it's good but they come here for the beaches. (DEN_TOU_06)

There is a strong orientation towards sustainability and a pervasive emphasis on ecological worth, with a central argument that Danish coastal environments need to be protected from overdevelopment and from mass tourism. This is particularly important as access to an undeveloped seascape is a major attractor for the core tourism market, which results in blending ecological and market orders of worth.

There is an urban/rural divide in the case of Denmark, with Copenhagen more focused on urban amenities (arts, local culture, history, cycling, food, and drink) and urban-coastal environments, while rural communities are focused on experiencing nature outside of the city. However, ecological worth applies to both urban and rural contexts. A tourism sector participant explains this by saying, "You go either to Copenhagen or you go to coastal Denmark, the rest of Denmark. There's been work put into this, using Copenhagen as a hub for getting tourists out in the countryside and it works better in some places. … But for the rest of Denmark, it's still not very developed" (DEN_TOU_17). However, despite its urban setting, Copenhagen can also be viewed as an urban eco-tourism destination. As Nilsson (2019) argues, much of Copenhagen's tourism image relies on bicycling, the clean city harbour environment, and a self-image of the city as a leader in sustainability and green innovation.

The Danish tourism sector is growing and is generally framed as providing social and economic benefits (market worth), enhancing the liveability of communities, increasing migration to host communities, increasing international visibility, and as providing opportunities to promote ecological literacy and environmental education. For example, a tourism participant notes, "The tourism industry is growing very fast and is very important for this part of Denmark. Because the people coming here, many of them are of good money. And stay here for several days. And that's very important" (DEN_TOU_O3). Another tourism sector participant similarly argues for the importance of tourism for rural communities, saying, "But as long as the tourists are still coming here, there's a basis for having shops in obscure places" (DEN_TOU_06). The notion that tourism is a source of pride for host communities also comes up for several of our participants, linked specifically to Wadden Sea National park, which receives international media visibility. There is also a sense that tourism helps make Copenhagen a more cosmopolitan and vibrant urban space.

Relative to other cases, our participants express minimal criticism of tourism development in Denmark. Where evident, criticism tends to focus on potential overdevelopment of coastal areas, negative impacts on wildlife, and the carbon intensity of travel, reinforcing the notion that ecological worth is dominant in Denmark. For example, a government participant notes the localized environmental impacts as follows:

> People say they're small things but maybe people go too close to the seals or they go to areas where the breeding birds are. … The normal behaviour of forgetting their own garbage and litter and I think those are the main aspects … I won't say the problems are big. But I definitely see problems there in the future. (DEN_GOV_12)

Whereas broadly positive views of tourism are prevalent, views of oil are more heterogeneous. Overall, there is a sense that the oil sector makes significant economic and employment contributions, especially in Esbjerg, which like Aberdeen was transformed from a fishery-based economy to an oil-based economy. An oil sector participant describes the process of Danish oil development as follows:

> So, the main impact of the oil and gas industry is that we, over the years, have secured welfare in Denmark. … And every one out of ten workplaces as Esbjerg as a city is linked to the oil and gas industry directly or indirectly. So obviously, it has a huge impact in that region. (DEN_OIL_08)

This transformation is heavily promoted by the municipality of Esbjerg, which self-identifies as Denmark's energy city, and the Esbjerg maritime museum, where oil and tourism appear as parallel development paths that need not conflict (fieldnotes, April 19, 2017).

These defences of oil invoke market worth, but again oil development in Denmark tends to clash with more salient conceptions of ecological worth, such as the widely shared notion that Denmark is a "green" country that is well into a renewable energy transition. For example, an oil sector participant describes this tension as follows:

> What we are about as an organization is that we get the most out of the Danish sector of the North Sea. But we realize as an organization that this has to be done in a way such that it creates maximum value for the Danish state and we have to take into consideration the societal impact, including CO2 emissions. (DEN_OIL_08)

Similarly, a government participant draws a comparison between Stavanger, Norway and Esbjerg, Denmark to illustrate the cultural differences in how oil is viewed in the two societies, reflecting the centrality of industrial and market worth in Norway and ecological worth in Denmark:

> The way people see Esbjerg in Denmark and the way people see Stavanger, for example, in Norway, are quite different. In Norway… people are very proud of the oil business. And so are we in Esbjerg, but if you look all across the Danish country, people are not so proud. They're not thinking so much that there's an oil and gas industry. They like that there is a lot of tax income and the government gets a lot of money. But in the end, they would like to reduce CO2 emissions and so on. (DEN_GOV_09)

While oil may coexist with renewables for the time being, the Danish continental shelf is in decline (i.e. post-peak) and is perceived by many in the public sphere as a sunset industry.

For the most part, Danish oil extraction is not seen as posing significant social or environmental risks, but is seen as responsibly managed, with no major episodes of catastrophe. The ecological risks of oil are rarely invoked by our participants. When they are, they focus on the possibility for catastrophic events, often invoking the 2010 BP Gulf of Mexico spill. Such catastrophic events are viewed as very unlikely in the Danish context. Where concerns are raised, they are specific and localized or are directed more at concern over the expansion of Arctic oil frontiers. For example, environmental organization concerns about ecological risks are more oriented towards Arctic oil exploration and Greenlandic oil development (field notes, May 9, 2017). These controversies are examined in greater depth in Chapter 5. As with other cases, we see a strong emphasis here on sustainability and a seemingly shared perception that risk mitigation is done well. However, the familiar tendency to respond to concerns grounded in ecological worth with appeals to solutions that have industrial worth is much less common in Denmark than in the other cases we have discussed so far.

Despite the marginalization of oil overall, there is a common understanding among many of our participants that tourism and oil development are in fact complementary, and this is often achieved by emphasizing market worth, and in some cases specific even specific mechanisms, such as spillover processes. For example, a tourism participant describes the coexistence of oil and tourism around Esbjerg and Wadden Sea National Park as follows:

> I think it's fantastic to be a part of Esbjerg municipalities [Wadden Sea National Park]. Because we are here, and we get money from Esbjerg municipalities to build the Wadden Sea Centre. Because they have money. And they have money from the oil industry. The oil industry is the biggest source of income in Esbjerg, that's without question. (DEN_TOU_05)

The oil sector drives regional migration, provides well-paying jobs, and contributes to affluence in a region, which spills over to the tourism sector. Conversely, the tourism sector prioritizes art, culture, food, and other things that make for more interesting and liveable places to settle, including for oil sector employees. Another government participant describes this dynamic by saying:

> In that way, indirectly, when you build up recreation infrastructure, outdoor facilities, you need money. And this is where the tax income from the oil sector can benefit the local area. … I know some green [environmental] organizations would say, 'oh, you're talking with the devil, you're playing with the devil, because they're polluting and doing everything.' But you say, okay, on the other hand, if nobody's paying for anything, how can we conserve nature. (DEN_GOV_12)

Denmark is already moving along the path to renewable energy transitions. There is a general sense that this is compatible with tourism development, in part because of the tight connections between tourism and ecological worth. For similar reasons, there is less of a sense that renewables are compatible with oil development, although some of our participants emphasized the possibility of coexistence.

Our Danish news story data is less informative than our news story data for other cases, which is likely due in part to the fact that our news corpus is smaller. The topic model contains a higher number of junk topics relative to other cases, which in this case could simply be an artefact of working with a smaller corpus. Still, we cautiously interpret the results as aligned with our findings from the qualitative analysis. The most dominant themes concern the Arctic and international relationships. We also see coherent themes forming around offshore oil extraction; marine transport; climate change; the economy and employment; and food, restaurants, and hotels. The hierarchical model, shown in Fig. 3.4, suggests that talk of oil intersects with environmental science, as well talk of a global Arctic, or representations of the Arctic as a space of global scientific, ecological, and political concern. The results also suggest that talk of climate change and the Arctic influences many other topics, including tourism development. The lower quality of these models, however, means that our interpretations should be viewed with measured scepticism.

In short, Denmark embraces tourism development more strongly than oil development, apart from the oil-dependent region of Esbjerg, which sits alongside Wadden Sea National Park. Ecological worth is unambiguously dominant, but both development models are seen as having positive social and economic impacts on host communities, and as we have now seen across nearly all cases, many participants draw on notions of market worth to cast them as complementary development paths.

Denmark closely resembles Iceland, discussed below, in that ecological worth is central and industrial worth is less common; both are also advanced in the transition to renewable energies and oil development is relatively marginalized.

Iceland

Like Demark, Iceland differs markedly from the other cases insofar as ecological worth dominates, oil development is barely a consideration, and industrial worth is less visible. Coastal communities and environments are predominantly portrayed in ways that are favourable to nature-oriented tourism, and which frame Iceland as a "green" country. For example, when asked what draws people to the region, a tourism

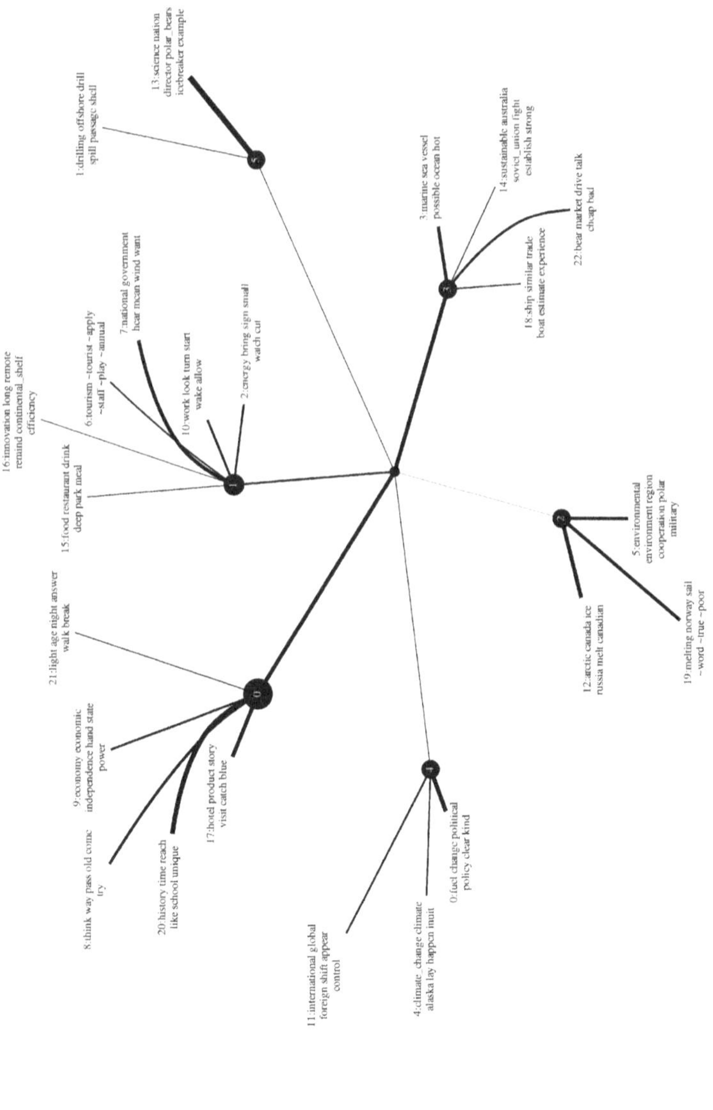

Fig. 3.4 Macro-level discourse network for Denmark

sector participant replies, "I think it's just the wild nature, the untouched nature that we have here, whether it's the wildlife around the country or just, you know, going to national parks or something like that" (ICE_TOU_23). Similarly, another tourism participant sketches out the range of natural, geological, and historic attractors as follows:

> If you go to a waterfall, if you can find that same waterfall with only ten people instead of five hundred it adds to the experience of nature without people. … I mean it kind of sells itself, you know. We have the glaciers—this is a whole set of things that are quite unique. And also, the historical aspect of the Vikings and all that. (ICE_TOU_18)

There is a strong emphasis on geysers, geothermal pools, the Northern Lights, volcanic landscapes, waterfalls, glaciers, and wildlife (especially whales and puffins), all of which can be experienced through activities like hiking and boat tours, and are framed in ways that are oriented towards inspirational worth.

Tourism is often framed through sustainability discourse and as a tool for environmental education, especially on marine ecologies. A tourism participant describes their practice as follows: "We received the Blue Flag this spring under new criteria, sustainable boating tourism. … It involves telling our customers about how we approach the animals, how we behave in the nature that we're going through" (ICE_TOU_23). The environmental costs of tourism, discussed below, are viewed as manageable. Rather, there is a common understanding that the benefits that environmental experiences and education tourism provide outweigh the ecological costs. Another tourism participant, for example, talks about using glacier tours as a forum for climate change awareness, saying:

> If you do a glacier walk with us, you're likely to have a guide speak about global warming. One of the interesting things is how the glaciers are changing with global warming. … We educate people, we make them interested in nature, we make them interested in preserving nature. (ICE_TOU_28)

Like the case of Scotland, talk about coastal environments and tourism is coupled with an emphasis on history (particularly Vikings), arts, food, and local culture as a repertoire of tourism attractors.

In addition to being positioned as environmentally sustainable, tourism development is seen as a mechanism for preserving history, providing social and economic benefits to host communities, and providing employment in rural communities. This blends ecological worth with market worth. For example, an ENGO participant states, "If we look at tourism and what it has brought to us after the financial crisis, it's basically helped us survive" (ICE_NGO_07). The same participant notes that tourism helps preserve history: "You have the national history museum here in Iceland, I think the main guests there are tourists" (ICE_NGO_07).

Despite being mostly positive, our participants also discuss the drawbacks and challenges associated with tourism development. Negative impacts include governing short-term rentals like AirBnB, dealing with rapid increases and decreases in the number of tourists, managing overcrowded tourist spaces, and the challenges of ensuring that the benefits of tourism spread across the whole country, not only in Reykjavik and the capital region where most tourism is concentrated. A government participant describes the trade-offs between positive and negative impacts as follows:

> The negative externality? We see warning signs of overcrowding in certain areas, where the percentage of tourists versus locals is getting a bit high … And it's this worrying sign that in the downtown area, as the same time as we're trying to get more buildings built there, more apartments are going to only AirBnB. (ICE_GOV_21)

There is reflexivity about both the local and extra-local environmental impacts of tourism, with a focus on ecological worth. At the local level, we see an emphasis on the negative environmental impacts of increased demands on infrastructure, overuse, and waste management. A tourism participant evokes the busy Golden Circle travel route to give a sense of environmental challenges:

> Take an area like the Golden Circle. We're not going to turn things around. You're never going to go into the glacier area and have the same experience that I had as a kid with my parents. What we have to do there is make sure it doesn't get damaged to the point where it's uninteresting. … But it comes down to the experience, protecting the nature and spreading out the traffic. (ICE_TOU_28)

Here, as in Norway and Scotland, there is an interest in mitigating these negative impacts with tourism governance and codes of conduct for tourists.

By contrast, there is limited talk of coastal environments as sites of oil prospectivity. Looking back at the earlier wave of exploration, an oil sector participant notes:

> There was this chance that there might be oil in a certain area in the north of Iceland, especially. The government decided to open up the continental shelf. And it has been a bit disappointing for the government and the people working on this. (ICE_OIL_07)

There are plans for a new shipping port at Finnafjord, which itself is premised on the notion of a changing Arctic and the increased openness and importance of Arctic shipping routes (Dodds 2010). As such, oil prospectivity is part of a nascent vision for extractive and shipping-oriented development in the northeast region of the country, which is outside the core tourist regions. Icelandic oil, however, requires a significant rebound of global oil prices that makes Arctic frontiers (i.e. northeast Iceland) more appealing, which seems increasingly unlikely. That said, some of our project data continues to convey optimism about a rebounding price cycle that will re-energize interest in Arctic oil prospectivity and bring potential revenues and employment to the northeast regions of the country, which are at the periphery of tourism development.

While energy sector data frames coastal prospectivity and extraction as having positive impacts for host communities, participants from the government, ENGO, and tourism sectors emphasize Iceland's wealth of renewable resources that make oil development unnecessary, and cite environment risks to wilderness values and ecological worth. For example, a government participant states:

> And it just goes without saying that if you're trying build upon, for example, this whale watching business here, how long could you fit with the impact of the industrial development here? And then the oil business? How can you manage that and make that harmony occur here in the harbour? It's a challenge. (ICE_GOV_22)

Another government participant notes similar tensions that make oil and tourism incompatible development paths: "If we become an oil nation it most likely won't benefit tourism. And it can be damaging to our image. Because we are selling Iceland as an environmentally clean country" (ICE_GOV_14). A few participants—all from government—argue that oil development can coexist with renewable energies, but in general this position reflects the dominance of renewable energy in Iceland more than the importance of oil development.

The promise of Icelandic oil has fallen out of the political/public imaginary and many see the pursuit of oil development as unnecessary given Iceland's wealth of renewable energy sources, mostly via hydropower plants and geothermal energy. More generally, oil development is seen as conflicting with Iceland's identity as a "green" society oriented towards ecological worth, with the Paris COP agreements and other responses to climate change, and with notions of a global Arctic. There is also a shared concern that oil development could harm Iceland's tourism image, furthering the sentiment that oil development and tourism development are incompatible.

Once again, these findings are reinforced by our semi-supervised topic models of Icelandic news stories. The news data is dominated by four themes focused on travel and mobilities, the Arctic and Icelandic nature, restaurants, history tourism, all of which are clustered together into one macro-level discourse, represented as cluster 1 in Fig. 3.5. Most of the focal concepts we used to seed this topic model do not produce coherent topics in Iceland, whereas they do in other cases. What is most notable, however, is that the only references to oil are future-oriented and are

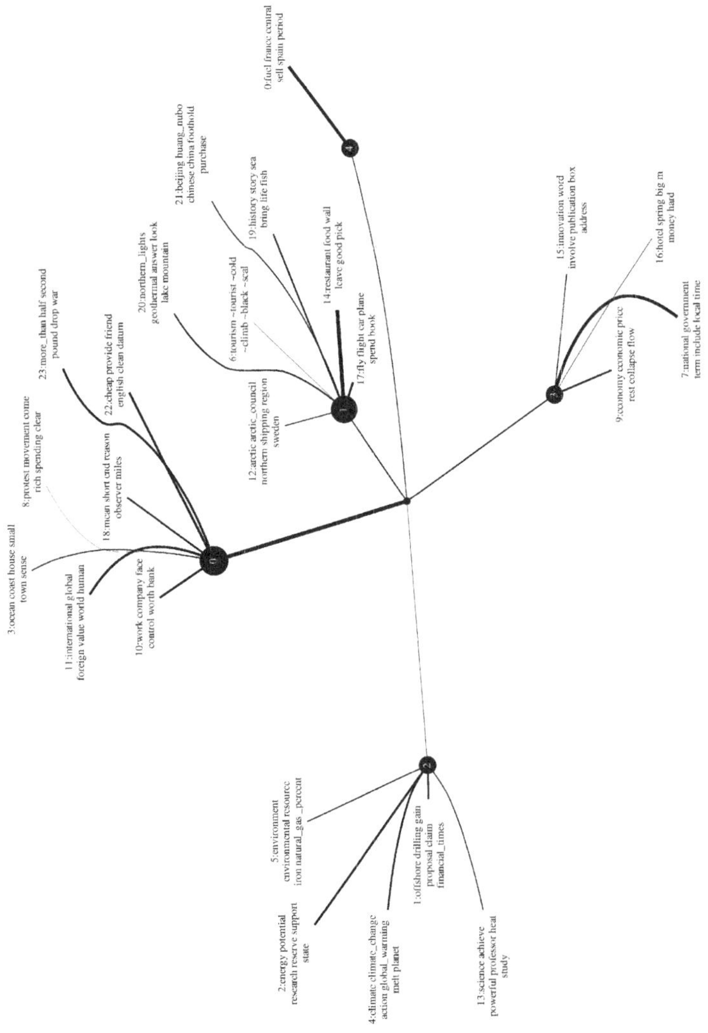

Fig. 3.5 Macro-level discourse network for Iceland

embedded in a discourse cluster (cluster 2) alongside topics related to climate change, climate science, and energy more broadly.

In sum, the cultural dimensions of oil-tourism interface in Iceland are heavily weighted in favour of nature-oriented tourism and, like Denmark, are anchored in the ecological order of worth. The challenges here involve managing the inevitable local environmental impacts of high volumes of tourism, but also the larger environmental costs of carbon-intensive travel. Oil development is hypothetical and future-oriented. Given recent declines in oil prices, conflicts over Arctic oil exploration, and Iceland's wealth of renewable energy sources, oil development is largely out of the public and political sphere. Unlike our other cases, oil and tourism are more often interpreted as incompatible development paths.

Conclusion

The goal of this chapter was to identify cultural variation in the oil-tourism interface across our five cases. To do so, we focused on how each development model is described and *evaluated* within and across cases. These cultural evaluations and concomitant boundary processes are embedded in institutionalized "orders of worth" that, by determining what is valuable and why, have real consequences for environmental governance and socio-ecological well-being. Across all cases, we see that nature-oriented inspirational worth is strongly present when it comes to general discussions of the environment. When it comes to oil and tourism development models, industrial, ecological, and market worth dominate, but the relative importance of each varies across cases.

We found that the criteria used in evaluations are generally homologous with the political economy framework introduced in Chapter 2. This suggests that cultural visibility, as an aspect of power, is broadly aligned with the relative political economic importance of the oil and tourism sectors in our case study regions. For example, Norway and Newfoundland and Labrador are both strongly oil oriented and, relative to our other cases, rely less on tourism development. Industrial and market orders of worth dominate; while ecological orders of worth are present, they tend to be folded into more general industrial worth, for example by emphasizing the role that scientific and technological innovation play in responses to climate change. To link this back to marine sociology, here we see how oceans are valued primarily as a natural resource pool (Hannigan 2016). Conversely, in the cases that are more tourism oriented and have less oil development—Iceland and Denmark—we find that ecological worth dominates, where oceans are valued as pathways of tourism and leisure travel (Urry 2014). Given that industrial worth tends to be more strongly associated with oil, it is perhaps unsurprising to see that market worth (e.g. the economic value of tourism) plays a more central role in these cases. Scotland differs from these other four cases, which reflects other differences that are grounded in the Scotland's political economy, bound up as it is in struggles around Brexit and independence, as well as issues of decommissioning old wells in a post-peak oil context.

We started this chapter by noting that these types of evaluations and boundary work have real consequences; evaluations are political and have power effects. As we have seen here, even when there are clear patterns within and across cases, there are inevitably going to be political clashes over what really matters, *what is valued and why*. In the coming chapters, we will see how evaluations like these, and the broader orders of worth they are embedded in, play out in arenas of environmental governance and collective action.

References

Blei, D. M. (2012). Probabilistic topic models. *Communications of the ACM, 55*(4), 77–84.

Boltanski, L., & Thévenot, L. (1999). The sociology of critical capacity. *European Journal of Social Theory, 2*(3), 359–377.

Boltanski, L., & Thévenot, L. (2006). *On justification: Economies of worth* (C. Porter, Trans.). Princeton, NJ: Princeton University Press.

Burkhus, W., & Ignatow., G. (2019). Cognitive sociology and the cultural mind: Debates, directions, and challenges. In W. Burkhus & G. Ignatow

(Eds.), *The Oxford handbook of cognitive sociology* (pp. 1–30). Oxford, UK: Oxford University Press.
DiMaggio, P., Nag, M., & Blei, D. (2013). Exploiting affinities between topic modeling and the sociological perspective on culture: Application to newspaper coverage of US government arts funding. *Poetics, 41*(6), 570–606.
Dodds, K. (2010). A polar Mediterranean? Accessibility, resources and sovereignty in the Arctic Ocean. *Global Policy, 1*(3), 303–311.
Emirbayer, M. (1997). Manifesto for a relational sociology. *American Journal of Sociology, 103*(2), 281–317.
Friedman, A. (2019). Cultural blind spots and blind fields: Collective forms of unawareness. In W. Burkhus & G. Ignatow (Eds.), *The Oxford handbook of cognitive sociology* (pp. 467–484). Oxford, UK: Oxford University Press.
Gallagher, R. J., Reing, K., Kale, D., & Ver Steeg, G. (2017). Anchored correlation explanation: Topic modeling with minimal domain knowledge. *Transactions of the Association for Computational Linguistics, 5,* 529–542.
Hannigan, J. (2016). *The geopolitics of deep oceans.* Cambridge, UK: Polity Press.
Kukkonen, A., Stoddart, M. C., & Ylä-Anttila, T. (2020). Actors and justifications in media debates on Arctic climate change in Finland and Canada: A network approach. *Acta Sociologica.* https://doi.org/10.1177/0001699319890902.
Lafaye, C., & Thévenot, L. (1993). Une justification écologique? Conflits dans l'aménagement de la nature. *Review Français de Sociologie, 34*(4), 495–524.
Lamont, M. (1992). *Money, morals, and manners: The culture of the French and the American upper-middle class.* Chicago: University of Chicago Press.
Lamont, M. (2012). Toward a comparative sociology of valuation and evaluation. *Annual Review of Sociology, 38,* 201–221.
Lamont, M., & Molnár, V. (2002). The study of boundaries in the social sciences. *Annual Review of Sociology, 28*(1), 167–195.
Lamont, M., Silva, G. M., Welburn, J., Guetzkow, J., Mizrachi, N., Herzog, H., et al. (2016). *Getting respect: Responding to stigma and discrimination in the United States, Brazil, and Israel.* Princeton, NJ: Princeton University Press.
Lamont, M., & Thévenot, L. (2000). *Rethinking comparative cultural sociology: Repertoires of evaluation in France and the United States.* Cambridge, UK: Cambridge University Press.

Leschziner, V. (2019). Dual process models in sociology. In W. Burkhus & G. Ignatow (Eds.), *The Oxford handbook of cognitive sociology* (pp. 169–191). Oxford, UK: Oxford University Press.

Light, R., & Cunningham, J. (2016). Oracles of peace: Topic modeling, cultural opportunity, and the nobel peace prize, 1902–2012. *Mobilization: An International Quarterly, 21*(1), 43–64.

McLevey, J. (2021). *Doing computational social science*. London: Sage.

Mohr, J. W., Bogdanov, P., DiMaggio, P., Nag, M., Blei, D., McFarland, D. A., et al. (2013). Topic models and the cultural sciences. *Poetics, 41*(6), 545–770.

Nelson, L. K. (2020). Computational grounded theory: A methodological framework. *Sociological Methods & Research, 49*(1), 3–42.

Nilsson, J. H. (2019). Urban bicycle tourism: Path dependencies and innovation in Greater Copenhagen. *Journal of Sustainable Tourism, 27*(11), 1648–1662.

Norgaard, K. M. (2011). *Living in denial: Climate change, emotions, and everyday life*. Cambridge, MA: MIT Press.

Roose, H., Roose, W., & Daenekindt, S. (2018). Trends in contemporary art discourse: Using topic models to analyze 25 years of professional art criticism. *Cultural Sociology, 12*(3), 303–324.

Scott, N. A. (2020). *Assembling moral mobilities: Cycling, cities, and the common good*. Lincoln: University of Nebraska Press.

Swidler, A. (1986). Culture in action: Symbols and strategies. *American Sociological Review, 51*(2), 273–286.

Swidler, A. (2013). *Talk of love: How culture matters*. Chicago: University of Chicago Press.

Thévenot, L., Moody, M., & Lafaye, C. (2000). Forms of valuing nature: Arguments and modes of justification in French and American environmental disputes. In M. Lamont & L. Thevenot (Eds.), *Rethinking comparative cultural sociology: Repertoires of evaluation in France and the United States* (pp. 229–272). Cambridge, UK: Cambridge University Press.

Thornton, P. H., & Ocasio, W. (2008). Institutional logics. In R. Greenwood, C. Oliver, K. Sahlin, & R. Suddaby (Eds.), *The Sage handbook of organizational institutionalism* (pp. 99–129). London: Sage.

Tilly, C. (2006). *Why? What happens when people give reasons … and why*. Princeton, NJ: Princeton University Press.

Urry, J. (2014). *Offshoring*. Cambridge, UK: Polity Press.

Vaisey, S. (2009). Motivation and justification: A dual-process model of culture in action. *American Journal of Sociology, 114*(6), 1675–1715.

Ylä-Anttila, T., & Kukkonen, A. (2014). How arguments are justified in the media debate on climate change in the USA and France. *International Journal of Innovation and Sustainable Development, 8*(4), 394–408.

Ylä-Anttila, T., & Luhtakallio, E. (2016). Justifications analysis: Understanding moral evaluations in public debates. *Sociological Research Online, 21*(4), 1–14.

4

Environmental Governance and the Oil-Tourism Interface

Building on the players and arenas framework, this chapter examines the political arenas where the work of environmental governance is carried out (Duyvendak and Jasper 2015). This perspective on government players and arenas compliments the environmental governance research that identifies two related shifts in environmental policymaking and management away from national governments as the only significant players. We can think of these as "horizontal" and "vertical" processes that shift how environmental governance is conceptualized and practised (Francesch-Huidobro 2012). Horizontal processes of governance occur when governments reach out and engage with a range of non-governmental players in environmental policy debate and decision-making. Vertical processes of governance occur when connections are made across international, national, regional, or municipal political arenas. In this chapter, we add to this conceptual framework by focusing on the connective arenas of governance, or the ways in which environmental governance creates arenas for engagement across environmental issues or policy arenas where different forms of development share social-ecological landscapes although they occupy different political jurisdictions or departmental silos.

M. C. J. Stoddart et al., *Industrial Development and Eco-Tourisms*,
https://doi.org/10.1007/978-3-030-55944-1_4

The horizontal expansion of environmental governance refers to the inclusion of a broader range of players into environmental policymaking, including environmental NGOs, scientists and experts, citizens, and business representatives. Research on horizontal governance focuses on how governments have structured the boundaries of non-governmental participation to be more or less inclusive, as well as how much influence non-governmental players are allowed in policy discussion and decision-making (Bach 2019; Mascarenhas and Scarce 2004). Researchers also assess if these horizontal processes create meaningful engagement or whether they simply provide the appearance of public participation while maintaining government control over processes and outcomes (Compas 2012; Montpetit 2003). In relation to resource extraction, for example, the ability of publics to meaningfully participate in environmental governance is often constrained by governments that prioritize the economic interests of oil or mining sectors (Dokis 2015; LeQuesne 2019; Parkins and Davidson 2008). From this perspective, non-governmental players gain power through their inclusion and influence in governance arenas. This can be accomplished through networks of collaboration and coalitions across individuals, groups, or sectors. Political efficacy can also be built by engaging in networks of conflict and contention (Castells 2009; Fischer and Sciarini 2015).

By contrast, the vertical expansion of environmental governance refers to the recognition that environmental planning and policymaking unfolds across multiple, intersecting scales of political jurisdiction. Complex environmental issues like climate change, fisheries management, or biodiversity protection involve collaboration and contestation between policymakers at the national scale, as well as within international political spheres organized through the United Nations, Arctic Council, or other multilateral bodies (Prior and Heinämäki 2017; Shadian 2014). At the same time, environmental policymaking also involves multi-scalar interaction between national and regional or local levels of government, which is complicated by power differentials among groups working at the different levels (Di Gregorio et al. 2019; Francesch-Huidobro 2012). Non-governmental players including NGOs, scientists, and corporate stakeholders may attempt to intervene at these various levels to advance their interests (Rootes 2013). Research on vertical processes of governance often focuses on the interplay between different scales. For example, researchers examine how debates and decision-making in arenas of global environmental governance are taken up by national governments, or conversely how the strategies of national governments support or impede the creation of international agreements (Lövbrand et al. 2017; Stoett 2012). Other work in this area focuses on initiatives at municipal or community scales and how these relate to national-level environmental policy objectives and practices (Bulkeley and Betsill 2005, 2013).

Research on horizontal and vertical governance processes attends to the range of players involved in environmental conflict, as well as the various political scales that intersect in grappling with environmental issues. However, focusing on horizontal and vertical dimensions of governance often misses the potential or actual connections between different environmental issues or forms of environmental management, which are often siloed in government policymaking and practice. To the environmental governance toolkit, we add the concept of connective arenas. These are specific arenas where links are made between forms of environmental management that are normally siloed to engage in more holistic planning and policymaking. Our concept of connective arenas parallels interest in landscape-level (or seascape-level) planning, as well as interest in cumulative effects of development, both of which encourage social-ecological systems thinking about of environmental policy and management (Barnes et al. 2017; Berkes 2015; Sayles et al. 2019; Schmidt et al. 2014; Young 2016). In the North Atlantic context, connective arenas of governance would create links across oil and tourism, as well as other vital sectors like fisheries, all of which rely on shared coastal and ocean environments.

After examining horizontal and vertical processes of environmental governance, we turn to connective arenas of environmental governance. We argue that the absence of connective arenas is the normal situation in environmental governance. This can be explained in part with reference to horizontal and vertical processes of environmental governance. Across most of our study regions, offshore oil governance involves a more concentrated network of governmental and national-international oil sector corporate players and is oriented around regional and national political arenas. By contrast, tourism governance involves a more diffuse and localized network of government and tourism sector players and is oriented around local or regional political arenas.

Horizontal Processes of Environmental Governance

Across cases, there is a great deal of oil-government contact and collaboration focused on oil exploration, environmental assessment processes, industry regulation, and promotion of host regions as sites of oil prospectivity and extraction. By contrast, tourism governance is less developed. Tourism development often involves collaboration between government and non-governmental actors around marketing and promotion. However, tourism governance could be strengthened because it often emerges in a reactive way after problems emerge and host societies begin to grapple with the costs of success. Figures 4.1 and 4.2 show constellations of key actors grouped by case study region for the oil sector (Fig. 4.1) and the tourism sector (Fig. 4.2).

We categorize horizontal governance in terms of levels of engagement. The Norwegian and Danish oil sectors, as well as the Danish and Scottish tourism sectors are relatively highly engaged on the horizontal dimension, featuring diverse groupings of well-integrated players, as well as state involvement that is proactive in cultivating resource extraction or tourism development for social wellbeing. The moderately engaged cases, such as the Scottish oil sector, or the Icelandic and Newfoundland and Labrador tourism sectors, have a more fragmented group of key actors, as well as state players that adopt more market-driven or reactive forms of engagement in resource extraction or tourism development. Finally, less engaged cases, such as the Icelandic oil sector or Norwegian tourism sector, are emergent or underdeveloped, where horizontal dimensions of governance are not well established related to either resource extraction or tourism development.

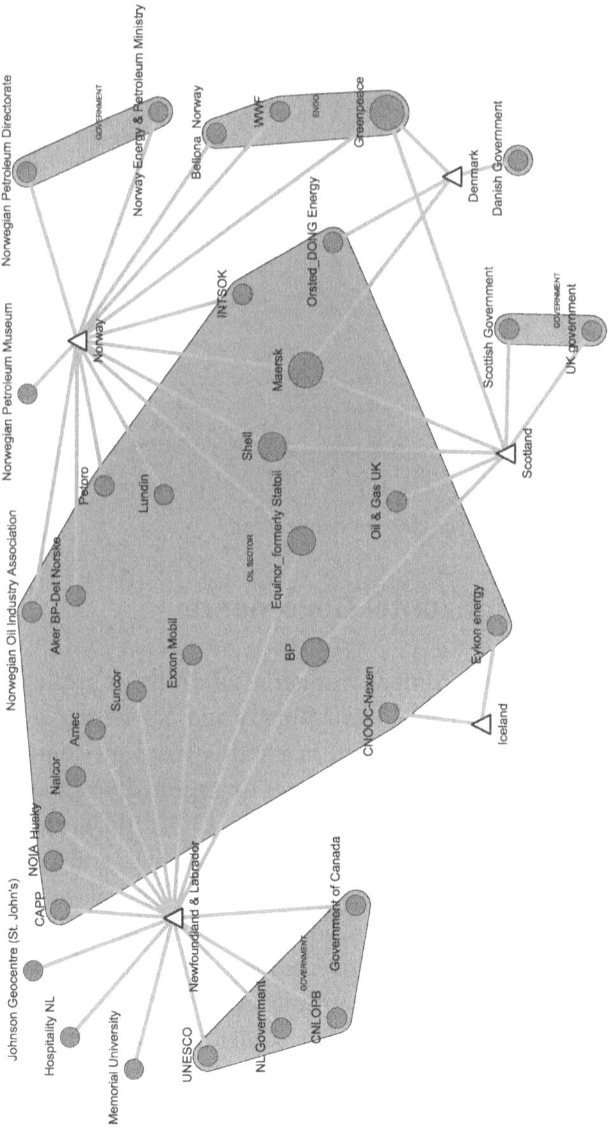

Fig. 4.1 Network of key actors for oil sector

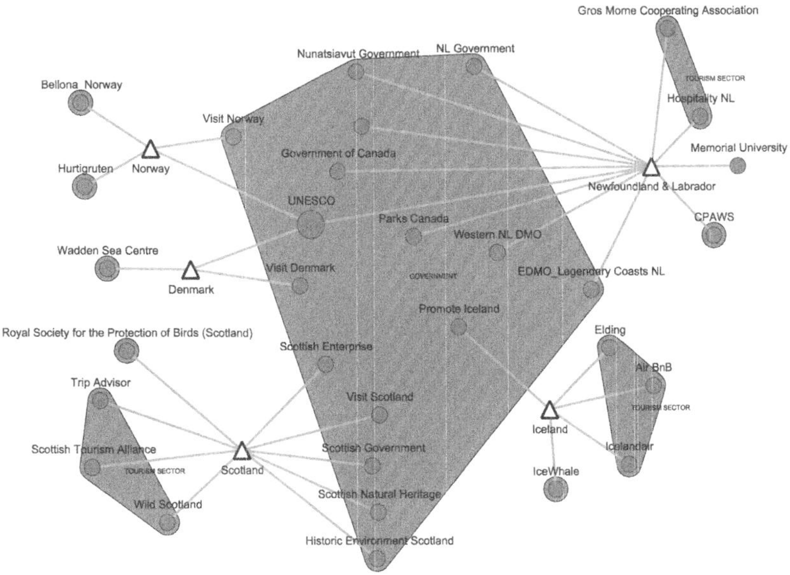

Fig. 4.2 Network of key actors for tourism sector

Highly Engaged Horizontal Governance

In Norway, national government ministries and offices are key actors, including the Energy and Petroleum Ministry and Norwegian Petroleum Directorate, as well as Petoro, which is a state-owned oil company (see Fig. 4.1). Equinor (formerly Statoil) comes up repeatedly as a key player, which is a hybrid public–private company. Other oil sector players include transnational companies like Maersk and Shell, and the Norwegian Oil Industry Association. INTSOK is also a key player, which is an organization that includes both government and oil sector partners and provides arenas dedicated for government-oil sector collaboration on research and innovation. Government-oil sector relationships are major factors in the success of the Norwegian model of oil development. This is consistent with other research that treats Norway as an exemplar of managing oil development to the benefit of host societies (Hilson and Newby 2015; Bryden 2015; Larsen 2006).

Environmental organizations also appear as players. This includes chapters of transnational groups like WWF, Greenpeace, and Friends of the Earth, as well as Norwegian organizations like the Bellona Foundation and Nature and Youth. ENGOs often conflict with the government and the oil sector over new oil exploration and extraction, such as the controversies in the Lofoten Islands region and around Arctic oil development. The Lofoten controversy is also examined by Kristoffersen and Dale (2014), who argue that localized collective identity and place attachment to Lofoten, with its fisheries-based culture and tourism sector, is important for understanding why opposition emerges in a country where oil is generally seen as the foundation of social wellbeing. Environmental organizations and research participants are sometimes critical of the "Norwegian fairy tale" and argue that the government is overly aligned with the interests of the oil sector. However, participants often note that the Norwegian political sphere is particularly open to civil society participation. An environmental organization participant explains this open political culture as follows:

> You probably heard it before, but in Norway it's quite open, there's easy access for us to talk to the government, there's easy access to talk to the political parties. The Parliament, for instance, it's easily accessible. But when you look at the different ministries, we have a lot of dialogue with the different ministries and the different directorates. .... We can talk with, more or less, whoever we want on any different topic. (NWY_NG0_04)

While there are points of tension and conflict, there is broad agreement that there is an open and collaborative political culture around oil governance (also see Ćetković and Skjærseth 2019). The Norwegian case shows a high degree of engagement in horizontal governance processes that integrate a broad range of players.

The Danish case is similar, with frequent invocations of the importance of oil-government collaboration. An oil sector participant notes, "The government understands that this industry contributes massively to the welfare [of Denmark]. The government understands the concept of the importance of security of supply. And the government understands that 34,000 workplaces are at stake. So, in that respect, there's a good collaboration" (DEN_OIL_08). Key energy sector players include Danish corporations like Ørsted (formerly DONG energy), Maersk, Ramboll Denmark, as well as transnational companies like Shell. Talk about collaboration often focuses on fiscal regimes that support oil development while creating economic and employment benefits for Denmark. The importance of government collaboration is also seen to promote environmental and social responsibility in the sector. Renewable energy transitions are another area of government-energy sector collaboration, with talk about how state engagement in the energy sector enabled Denmark to take an early leadership role in wind power transitions. An oil sector participant describes this process and the current state of the Danish energy sector as follows:

> I think it's fair to say that if you look at the energy industry as a whole, the focus in Denmark has been on the green energy transition for at least twenty years. And there has been a debate, recently, on whether it's correct that government is still supporting [oil] exploration and production activities in the North Sea. The vast majority of parties in the parliament are still supporting this. It's not either a green transition or oil and gas, it's both. Maybe even the oil and gas sector, the income from that can support the green transition. Some politicians are looking at it that way. So that's the broad context if you look at the oil and gas sector. (DEN_OIL_01)

Environmental groups, particularly Greenpeace and WWF, come up as players in oil governance. However, most environmental organizations conflict with the oil sector and government focuses on international issues, particularly on the extension of Arctic oil frontiers, rather than on operations in the Danish North Sea, where the oil sector has been established since the 1970s.

Tourism-government collaboration is also prevalent in the Danish case, with Visit Denmark appearing as a player that bridges tourism operators and government. As a tourism sector participant describes this relationship, "We depend on collaboration because we are a national organization. We provide some funding, but we also initiate projects, but the projects should be operated, have to be operated through our partners. And that would be the municipalities and the local DMOs [Destination Management Organizations]. And what we deal with is business development primarily" (DEN_TOU_18). Later, the same participant describes relationships between their organization and government by saying, "Well we've got very positive dialogue both with the national governments and the regional governments. Of course, it has to be a dialogue" (DEN_TOU_18).

There are also tensions related to tourism governance, particularly around ensuring the protection of Danish coastlines. Tensions around coastal zone development reflect competing visions for the kind of tourism that is appropriate for coastal environments, which a tourism sector participant describes as follows:

> Danish legislation has been successful in protecting the coast from large buildings, large hotels. You don't find large hotels here; you find a lot of summer houses. But almost no large hotels. And that's unique for Denmark. It's a feature which a lot of people in Denmark cherish very much and are willing to go a long way to protect that feature. But being such an industry that you want to develop, politicians are very tempted to let go of this legislation and sort of make it more investment popular. … But I think they should really be careful not to destroy the main asset, which is the beaches and the dunes. (DEN_TOU_06)

There are conflicting visions for how coastal zone management should be structured to juggle environmental protection and tourism development. Alongside this, however, we see tourism governance that is highly engaged across government and tourism sector players.

Scotland demonstrates high levels of horizontal engagement between government and the tourism sector. Government players include Visit Scotland, Scottish Enterprise, Historic Environment Scotland, Scottish Natural Heritage, and the Scottish Government. Tourism sector players include the Scottish Tourism Alliance and Wild Scotland. The NGO the Royal Society for the Protection of Birds (Scotland) also appears as a player involved in tourism development. Government-tourism collaboration focuses on marketing and promotion, but also on strategy and tourism management. This collaborative relationship is viewed as something that evolved in recent years. It shifted from a top-down government-led relationship to more of a bottom-up, industry-led collaborative relationship. This is described by a government participant as follows:

> And the approach in terms of tourism strategy prior to 2012, it was essentially a government strategy. Government strategies have their own merits. But the deliberate move was made that in 2012, the Scottish Tourism Alliance … would be given the task of an industry-based, industry-led strategy. Which they did. We provided analytical support, a huge amount of analytical support. And what happened was that the industry base, they chose what its targets were going to be. So, the industry vision was for sustainable growth and an excellent customer experience. And the key performance indicator was going to be looking at tourism revenue. Not necessarily numbers, but tourism revenue. (SCO_GOV_02)

Tourism sector participants similarly note an improvement in tourism-government collaboration and tie this to the 2007 shift to an SNP (Scottish National Party) government. The current relationship between tourism umbrella organizations and government is described as follows, "Visit Scotland, along with Scottish Enterprise, Highlands and Islands Enterprise … we provide experts, skills and knowledge to undertake the development of the sectors we're responsible for. Which then frees up the government to worry more about policy and direction" (SCO_TOU_01). Likewise, there is a high level of contact within the sector on issues of tourism promotion and development through players like the Scottish Tourism Alliance and Visit Scotland.

Moderately Engaged Horizontal Governance

In Scotland, government collaboration with the oil sector involves key players including the Scottish Government, Scottish Enterprise, the Oil & Gas Authority UK, and UK Government. Oil & Gas UK, the industry association body, also appears as a key player. Government-oil collaboration focuses on policy/fiscal regimes, risk mitigation, and supporting research and development. Whereas the Nordic cases have a greater emphasis on leveraging oil wealth for social development, the Scottish case orients around market-driven logics. When asked about oil sector relationships with government, an oil sector says:

> The relationship has largely been focused on taxation, in my humble opinion, historically. And that's still important, of course. And there's been quite a lot of tax changes. In fact, the last three or four years [there has been] a new tax regime, because of the falling oil price the government changes its approaches. And of course, the tax regime is global — lots of different countries offering their own tax regime. People that know better than me tell me that the UK is now one of the most fiscally attractive places for oil and gas. So that's good from an investment perspective. (SCO_OIL_19)

The same participant continues to describe how oil and government collaboration is also expanding into research and development. Horizontal processes of oil governance also focus on future-oriented issues including renewable energy development and decommissioning of platforms in post-peak operating areas. The focus on decommissioning is characteristic of the Scottish case, and is viewed as a key task in thinking about energy futures (McCauley 2018). Participants from oil and government share a recognition of the importance of decommissioning, but there is contention about which players should be responsible for carrying out and paying the cost of decommissioning infrastructure from post-peak fields.

Newfoundland and Labrador includes a substantial focus on oil-government collaboration. Key players include government agencies like the Canada-Newfoundland and Labrador Offshore Petroleum Board (CNLOPB), which is a joint body of the national and provincial governments, as well as the Newfoundland and Labrador Government (particularly the Ministry of Natural Resources). Significant oil sector players include transnational companies like Equinor, BP and Exxon Mobil; companies that are Canadian focused, such as Husky and Suncor; as well as Nalcor, the provincial public company focused on exploration. Other players include the Canadian Association of Petroleum Producers and the Newfoundland and Labrador Offshore Oil and Gas Industries Association (NOIA). At events like NOIA (the annual oil and gas industry conferences), talk about government-oil sector collaboration is often linked with positive views of the oil sector as a field of technological innovation. There is also a recurrent market-oriented discourse that government regulations cannot be too complex or onerous to attract and maintain investment. When asked to describe the oil-government relationship in Newfoundland and Labrador, an oil sector participant replies that:

> It has to be nurtured over time. Because there will be some differences in terms of the right level of regulation and the right concerns regarding impacts. I mean, from the oil industry perspective, you know, they're going to want to have the ability to undertake their business in a responsible way without being over managed by the regulatory sector. And from the regulatory point of view, they're going to want the industry to be responsible. And when the two wavers from the ideal, there will be some interaction that hopefully leads to correction, both sides. … And I think that generally over time, they've found a path that works in the interest of both. (NL_OIL_04)

As in other cases, the positive view of oil governance reflects the perception that the oil economy makes significant contributions to social wellbeing through government revenues and employment (Dodd 2012). While there is a role for public engagement in oil governance, there is less visibility for environmental movement players than in other cases.

Newfoundland and Labrador also shows significant evidence of tourism-government collaboration, primarily around promotion and marketing. Much of the collaboration works through provincial government players, such as the provincial Department of Tourism, Culture, Industry, and Innovation, as well as Destination Management Organizations. Parks Canada, a national government agency, also appears as a key player. The tourism sector peak organization, Hospitality NL, is the most visible tourism-oriented organization. The Gros Morne Cooperating Association also comes up repeatedly as a key player, though their focus is on collaboration and tourism development in Gros Morne National Park. The NGO the Canadian Parks and Wilderness Society (CPAWS) also works as a collaborator on geotourism development projects, where tourism is explicitly linked to environmental education and sustainability goals, as well as intervening to protect tourism landscapes from oil development around Gros Morne and the Gulf of St. Lawrence. In contrast to oil governance, however, fewer influential ENGOs intervene in tourism governance.

One characteristic that sets the Newfoundland and Labrador case apart is the greater visibility of Indigenous organizations, such as the Nunatsiavut Government, in tourism governance. The Nunatsiavut website emphasizes their engagement in the co-management of Torngats National Park in northern Labrador, as well as Inuit participation in research, training, and employment in the park. The 2018 Hospitality NL conference also reflected the increasing involvement of Indigenous communities from Labrador (Nunatsiavut, NunaKuvaut, Innu) and the island of Newfoundland (Qalipu and Conne River Mi'kmaq communities) in tourism as a strategy for community development.

Iceland demonstrates the need to build tourism governance to grapple with the challenges of success. Government-tourism sector collaboration is visible around tourism marketing and promotion, with Promote Iceland as a key governmental player. The tourism sector is dominated by a relatively small number of big players, such as Icelandair, the Blue Lagoon, and Elding, which have the capacity to engage in tourism governance and collaboration. Going beyond tourism promotion, however, there is a narrative of tourism fatigue and a need to respond to the emerging problems of overtourism. When asked about government-tourism collaboration, a local government participant responds, "Informally, a lot. But I think we can do a much better job in the formal setting." They further describe the need to develop tourism governance to better manage local tourism development as follows:

> I truly believe that it's a fact that without the tourism industry we would be in incredible difficulties in running a proper community here. Certainly, it has become one of the pillars of the local economy. … We need to make sure that we have both the foundations and the proper environment for these companies to thrive. But it needs to be a dialogue. And we need to know where we're going, we need to have a clear vision of how we want things to develop … The municipalities need to and certainly can do better in organizing meetings, organizing the dialogue

> that needs to be between the municipality and these companies, to maybe foresee certain things that may be at the horizon in two years. So, we can respond in time and at least do something to meet the expectations of all these companies. (ICE_GOV_22)

There are also signs of collaboration between tourism actors, government, and environmental groups around environmental sustainability and the protection of natural areas, industry self-regulation (such as the VAKINN eco-certification program), and boat tours.

Opposition to Icelandic whaling has been another issue where tourism interests and environmental players have aligned. IceWhale, in particular, is a player that leverages tourism to shift Icelandic environmental policy and to redefine whales from consumable nature into nature as object of touristic experience (also see Cunningham et al. 2012). An NGO participant describes the evolution of this collaborative relationship:

> They [boat tour operators] were worried about the impact [whaling] might have on their operations and people's perceptions of whale watching in Iceland. So, in some ways we can say we can thank whale watching for this operation because it is quite unique to get competitors to work together. And though we haven't been able to stop the whaling we have been able to do things otherwise that otherwise we might not have been able to. For example, developed a code of conduct for responsible whale watching, since Iceland has no regulation or law for that. (ICE_NGO_07)

Here, we see how tourism collaboration with environmental organizations can impact tourism governance and address gaps in formal, government-led processes of environmental governance.

Less Engaged Horizontal Governance

Iceland is an outlier compared with our other case studies, as we see less attention to oil governance, which is not surprising because exploration activity in the Dragon (Dreki) region has cooled significantly. The key corporate players that come up are CNOOC-Nexen and Eykon Energy, which link Chinese oil sector interests in Arctic oil development with players from the Icelandic energy sector. The Icelandic National Energy Authority also comes up as a government player. However, there is a sense that oil exploration is now under the radar of public and political discussion. When issues of oil governance arise, there is some talk about engagement by opposition parties and environmental groups, who raise questions about whether Iceland should pursue oil development. An environmental organization participant notes, "We haven't made a formal opinion this [oil development]. But I think we agree ... that this is something to worry about, especially if we look at the impact of the sound pollution. And how that could be affecting the whale watching up north, especially" (ICE_NGO_07).

Finally, the Norwegian case shows less of an emphasis on tourism governance. In contrast to oil, which has a high degree of visibility in the national political sphere, tourism is localized or regionalized in terms of governance, with the notable exception of Visit Norway, the national tourism promotion organization. Recurrent themes include that tourism governance is underdeveloped and that tourism governance is a local issue. Tourism is often interpreted as less of a political priority and as a result is more diffuse and fragmented. However, regions of Norway receive significant tourism flows, including fjord landscapes of the central coast and hub communities like Bergen (Oklevik et al. 2019). The Lofoten islands are notable as a nature-oriented tourism landscape. Tourism numbers have increased in recent years, which is creating problems of success. This is provoking communities in the Lofoten Islands to rethink tourism governance to mitigate the negative social and ecological impacts of tourism. For example, a local government participant describes the situation as follows: "It [the tourism boom] just started. I would say that we saw this coming, say, briefly, last year or the year before. But it's this year that we are really facing this and have to start planning. There's something we need to do before next season" (NWY_GOV_16). The same participant observes that conversations about "a strategy for what kind of guests we want to have visit the Islands" is taking place locally, with tensions around the desirability of cruise ships and increasing RV (or caravan) traffic. However, solving issues of tourism governance in Lofoten requires getting the attention of national policymakers. This participant is optimistic about the potential for building governance to address the negative impacts of tourism, concluding, "Hopefully this will be a national political issue in maybe a year or two" (NWY_GOV_16).

Vertical Processes of Environmental Governance

Across our cases, oil governance orients around the national and international scales, whereas tourism governance operates more at regional or local scales. The oil sector is an international economic network and several companies operate across multiple case study sites, including: Equinor, Maersk, Shell, and BP. Tourism governance is more diffuse, but it involves managing connections between international flows of travellers and local host communities. In terms of vertical environmental governance processes, we categorize the cases as predominantly oriented around international-national processes, national-regional processes, and regional-local processes.

International-National Vertical Governance

In Norway, oil governance is central to the national political sphere. It is within the purview of the national government and oil issues are subjects of national public and political debate. An environmental organization participant puts it simply: "Oil drilling is national politics" (NWY_NGO_02). The major state players are various national government offices and agencies (see Fig. 4.1). The narrative of Norwegian oil development constructed by the Stavanger oil museum similarly ascribes a central role to the national government in adopting a strong stance in relation to the oil sector and structuring oil development to maximize economic and social wellbeing for the public.

The 2015 Paris Agreement is also invoked as an international critical event that impacts Norwegian debates about oil, social futures, and decarbonization. For example, the Paris agreement came up repeatedly during fieldwork at the 2016 Offshore North Seas oil sector conference. A representative of the International Energy Agency noted that the Paris COP agreement presents a "united world" regarding the importance of climate change and significant political change, which demands adaptation and transformation by the oil sector (field notes, August 31, 2016).

In the Danish case, oil governance also orients around the national political sphere. The Danish national government is the main state player in oil governance. Danish-based Ørsted (formerly DONG Energy) and Maersk are also key energy sector players (see Fig. 4.1). National oil governance processes focus on issues such as fiscal and tax regimes for the industry, worker safety, and environmental assessment. At the same time there is a stated commitment at the national level to end domestic fossil fuel consumption by 2050. As an oil sector participant puts it:

> The political plan is to be independent of fossil fuels in Denmark by 2050. Not meaning that we are not allowed to produce oil in 2050. But we should produce more renewables than the energy we are using. But then we could export our own oil and gas if we are still producing at that time. I guess we can still produce oil and gas until 2050. If you look at it from the technical point of view, whether it's economical very much depends on the oil price. (DEN_OIL_01)

As this respondent highlights, the future of the Danish oil sector is configured by the interface of national politics and international flows of energy production and consumption. Even as Denmark shifts away from fossil fuel consumption domestically, ongoing oil extraction can persist into the future if there are international markets for oil as part of the global energy system.

The 2015 Paris agreement and UN climate governance arena also provoke a need to reconsider the Danish oil sector. When asked about oil and environmental sustainability, an oil industry participant frames their response through the Paris agreement by saying, "Even if you look at the COP21 scenario, with the two percent… reduction of the global temperature to 2 degrees. Even in that scenario, the International Energy Agency expects that 44 percent of world energy consumption will be oil and gas" (DEN_OIL_08). They go on to note that the Danish population has become accustomed to a high standard of living so it is untenable to just "close down the North Sea." Instead, they continue, "So what do we need to do? I think we need to focus in on bringing down CO_2 emissions so that it's not a fossil free society but a CO_2 emission neutral society" (DEN_OIL_08). Elsewhere, NGOs like Concito invoke the Paris Agreement and COP process to assert the need to integrate national climate policy into concerted global action through emissions targets and new energy sector technologies including carbon capture and storage.

On the tourism side of the Danish case, the national tourism agency Visit Denmark is a central player. Their "objective is to get tourists from the world to Denmark" (DEN_GOV_11). They also play an important role in providing "knowledge, analysis and surveys nationally and internationally" (DEN_GOV_12). Wadden Sea National Park, located on the west coast in Jutland, is also a space of multilevel processes that bridge from the international to the local scales. The creation of the park and its governance involve participation by local communities, local tourism operators, national government, and UNESCO. Wadden Sea National Park is also part of an international network of protected areas that reaches Germany and the Netherlands. A government participant describes the formation of the park as follows: "The Danish, Dutch and German government had collaboration on the Wadden Sea area in a broad sense. And the national park is born of those international responsibilities, together with the [UNESCO] World Heritage ... So, we are born into an international cooperation" (DEN_GOV_12). The National Park is defined as a local space of environmental education and protection but is also positioned in a global network of ecologically significant sites based on bird migration routes and demarcated by its UNESCO status.

In the case of Icelandic tourism governance, key players operate predominantly at either the national or local scales. Tourism sector players that come up repeatedly include large organizations like Icelandair, the Icelandic Travel Industry Association, and Promote Iceland, as well as more localized operators like the Blue Lagoon and a variety of whale watching operators. There is, however, a sense of weak connectivity across the national and local scales, with recurrent themes including: tourism governance is underdeveloped, tourism governance is seen as a local issue, and tourism governance is not a national political priority. As a government participant describes the recent tourism boom, "Even though tourism has grown as much as it has, we are not having that much impact on the political spectrum as we thought we might. Because the government is still thinking 'fisheries and agriculture'" (ICE_GOV_14). This leads to talk about the need for greater centralization and oversight of tourism planning and development at the national level and for the national government to take a more active role in tourism governance to ease the strain of the tourism boom on community infrastructure. However, there are also signs that tourism is becoming more of a national priority and that tourism governance is evolving.

Icelandic tourism also connects to international governance through initiatives like the European Union's Blue Flag program. This is brought up by a tourism participant to describe sustainability practices and eco-certification in relation to boat tours and whale watching. They say, "We received the Blue Flag this spring under new criteria this spring, sustainable boating tourism. … We're one of only four companies in the world to have this, the blue flag under this criterion." When prompted to explain this in more detail, they respond, "It's a long process to go through. … It involves telling our customers about how we approach the animals, how we behave in the nature that we're going through. Recycling, of course. Reducing. … it's a new criterion by the Blue Flag. These four [boat tour] companies are all actually in Iceland" (ICE_TOU_23).

Icelandic oil governance also orients around the national and international vertical scales, though oil exploration activity has gone dormant since recent global price declines and volatility. The main players that come up in relation to oil development are national governmental players, particularly the Iceland National Energy Agency. Key energy sector players are based in Iceland and internationally, including Eykon Energy and CNOOC-Nexen. As such, if Icelandic oil exploration re-emerges, then oil governance will likely assume a national–international orientation.

National-Regional Vertical Governance

In the Scottish case, key governmental and oil sector players operate in UK and Scottish political arenas, including the UK Government, Oil & Gas Authority UK, Oil & Gas UK, the Scottish Government, and Scottish Enterprise. Oil governance clusters around the Scottish government, though there are tensions in the relationship between the Scottish and UK governments. For example, when asked about oil sector-government collaboration, an oil sector participant notes, "You have the different government groups. You have the Scottish government and you have the UK government. And you have to find which ones to talk to" (SCO_OIL_02). Vertical tensions emerge in relation to oil as a source of national revenues and economic wellbeing, whereas there are perceptions that the city of Aberdeen, as a host city, has not been particularly successful at capturing social benefits of oil development for the city. The vertical dimension of governance is shaped by overlapping tensions between Scotland and the United Kingdom, including around potential Scottish independence, the Brexit process, and Scotland's relationship with the European Union. Brexit came up repeatedly during the Oil & Gas UK event that launched their 2017 annual outlook report. Participants noted that Brexit will create impacts on the industry in terms of taxes, jobs, innovation, and technology, but whether the net effect is negative or positive is unpredictable (field notes, March 7, 2017).

In Newfoundland and Labrador, oil governance operates jointly across the provincial and national scales, with the Canada-Newfoundland and Labrador Offshore Petroleum Board (CNLOPB) serving as a key player that bridges these political scales. Joint management creates tension, which comes up in fieldwork at the 2016 and 2017 NOIA (Newfoundland and Labrador Offshore Oil and Gas Industries Association) annual conferences. Conference speakers from the provincial government, including the Premier, raised concerns about proposed changes to national Environmental Assessment regulations that increase "regulatory uncertainty" for oil sector players and risk making the province less desirable as space for oil extraction (field notes June 20, 2017). From this perspective, any move by the Government of Canada to increase environmental oversight poses risks to the economic wellbeing of Newfoundland and Labrador. Conversely, the provincial Minister of Natural Resources argues for the need to "modernize regulations and improve governance" in order to allow more "nimble" assessment processes that support easier "frontier access" for new oil exploration and development (field notes June 23, 2016).

Regional-Local Vertical Governance

In Newfoundland and Labrador, while oil governance operates at the provincial and national levels, tourism governance orients towards provincial, regional, and local levels of governance. Governmental players include the provincial Department of Culture, Tourism, Industry and Innovation, and the various regional Destination Management Organizations. On the tourism sector side, key players include Hospitality NL, which provides a collective voice for the provincial tourism sector. The regionalized nature of tourism governance is described by a government participant who contrasts it with the political weight of resource extraction as follows, "From an industry, leadership and industry association, HNL [Hospitality NL], the DMO [Destination Management Organizations] mechanisms – even within provincial departments, I would say tourism, it's not only second or third or fifth or sixteenth. They definitely take a backseat to [resource extraction sectors]" (NL_GOV_07). Notable exceptions to this local and regional focus include National Parks and UNESCO sites, such as Gros Morne National Park, which are spaces of multilevel governance related to tourism. Here, international (UNESCO) and national (Parks Canada) players are more prominent and active alongside provincial and local players.

In our Scottish data, the main tourism governance players are Scottish governmental and tourism sector agencies including Visit Scotland, the Scottish Government, Scottish Natural Heritage, Historic Environment Scotland, and Wild Scotland. In contrast with oil governance, there is limited visibility for UK governmental players. The importance of collaboration across regional and local scales, and active involvement by local communities in tourism development is emphasized. This comes up in relation to managing some of the challenges with tourism around infrastructure development and maintenance. As a government participant notes, "We can have any amount of [strategies] we want and we can align plans, but at some point, you're talking about tarmac, you're talking about toilets and you're talking about a cup of tea. Those have always been the holy trinity of what tourists want. And again, it's the pressures on what happens when they're not supplied or when a smaller community [needs] these things" (SCO_GOV_02). However, anxieties around Brexit also arise about uncertainty for the tourism labour force, which is highly dependent on EU nationals (field notes, March 22, 2017). Anxiety over Brexit demonstrates how actions in the national political arena impacts Scottish tourism development at the regional and local scales. Tourism operators are looking to groups like the Scottish Tourism Alliance as a collective voice to help navigate this uncertainty.

In the Norwegian case, tourism governance is peripheral to the national political sphere, though the national tourism organization Visit Norway is as a key player. As indicated on their website, Visit Norway promotes tourism as a sustainable development pathway and coordinates eco-labelling and sustainability initiatives. However, as a tourism sector participant observes, "You have several thousand small travel businesses that don't necessarily cooperate" (NWY_TOU_19). Instead, the challenges of tourism are more often relegated to municipal and local political spheres to manage. A government participant describes localized challenges to building tourism governance at Lofoten as follows:

> All the hotels, all the camping, all the tent areas – every place you can pay to spend the night has been fully booked the last three or four years. So, the bad things that are happening it's mainly because we don't have enough numbers of tenting areas, of camping places for the car loans [rentals] and we don't have more hotel rooms. But, also, we have another type of people coming here, they want to see the wild, the untouched. ... We have to make some changes to take care of the issues that we are facing. If we can do that, we don't have problems with the locals, we don't have problems with the ones who are living off the tourists, we don't have problems with the shops. It's much easier to have control. (NWY_GOV_19)

Several participants raise the possibility of a tourist tax as a tool for providing capacity for host communities to respond to the localized negative impacts of tourism, including overcrowding of tourism spaces and travel infrastructure, as well as waste management issues. However, the idea of a tourist tax is also contentious and requires scale bridging beyond the local level to implement. The localized nature of tourism governance also means that there is a scale mismatch that works against developing conversations across oil and tourism sectors in Norway.

Connective Arenas for Environmental Governance

Connective arenas of governance are spaces for offshore oil, tourism, and other players to come into contact to shape landscape or seascape-level visions for coastal development. Offshore oil and tourism are often viewed as complimentary development paths. However, connective arenas are weakly developed across our case studies. Furthermore, they tend to be limited to points of conflict between oil and tourism interests over specific projects. We can break down oil-tourism connectivity into three types: complimentary but parallel development paths; conflictual connective arenas; and disconnected development paths.

Complimentary but Parallel Development Paths

The first type of connectivity interprets both tourism and oil development as beneficial for host societies, with little interaction between the sectors. In the Danish case, there is little sense of connectivity between tourism and oil development. While the oil sector is seen as having positive impacts for host communities, oil is often described as important to specific regions of the country. For example, Esbjerg brands itself as an "Energy Metropolis" in reference to its history as an oil hub city, as well as an entry point for offshore wind energy. As an oil sector interviewee notes, "And every one out of ten workplaces as Esbjerg as a city is linked to the oil and gas industry directly or indirectly. So obviously, it has a huge impact in that region" (DEN_OIL_O8). However, a recurring theme describes a social-political divide between the capital of Copenhagen, with its image as a tourism hub and green city, and much of the rest of the country. As another oil sector participant puts it, "It's a shame. But on the other hand, a lot of Danes think that this sector is rubbish. That we should close down. Because there is no future in it. The future rests with the renewables" (DEN_OIL_01).

The west coast of Jutland provides an example of a landscape shared by offshore oil, oriented around the community of Esbjerg, and nature-based tourism, oriented around Wadden Sea National Park. As one government participant notes, both oil and tourism are "vital" for this region (DEN_GOV_12). Another government participant expresses this as follows:

> Now, you could say that there should be a conflict between the [oil] industry and the local area because we are a national park. Just outside our harbour, a national park starts called the Wadden Sea. And now two or three years ago we were put on the list of the World Heritage sites. And you could say that they ought to be conflicts between these two things [oil and tourism] but I don't think there is a conflict. I think we've found out how to live as good neighbours. (DEN_GOV_09)

Oil and tourism are seen as complimentary development paths, with some indirect contact across these sectors that share social and ecological space. A participant notes the oil sector in Esbjerg creates a "positive spiral" because it gives an economic foundation and "you can attract other people there; they will attract other people. Then you've got a different form of city development and that, again, attracts tourists. So, I would argue that it's a spill over effect" (DEN_OIL_08). A government respondent similarly notes indirect connections as follows: "When you build up recreation infrastructure, outdoor facilities, you need money. And this is where the tax income from the oil sector can benefit the local area" (DEN_GOV_12). In this landscape, where both oil and tourism are valued, there is little evidence of formalized relationships through governance processes, with an oil sector participant noting that the two industries have "similar initiatives" but little direct interaction

(DEN_OIL_08). However, there is evidence of informal connectivity. For example, the Esbjerg Maritime Museum devotes significant exhibit space to the history and social-cultural importance of oil development in the region. The museum receives resources and informal input from oil companies. Elsewhere, stakeholders from both oil and tourism players engage in conversations about business and regional development, for example in regional working groups. Here, we see examples of nascent or subsurface connective arenas.

In Scotland, both tourism and oil are prominent in the public imaginary and in political arenas. They are largely seen as complementary development paths, with both providing positive economic impacts for host communities. A government participant notes the spillover economic effects for host communities like Shetland as follows:

> There has been some very interesting stuff in terms of the years up until now [before the oil price downturn]. When you've seen, particularly, oil funds used in some of our islands, you know, you have a huge set of infrastructures. Perhaps, you might say, Shetland has more swimming pools per head of population or whatever … you get that perception, that they've used them [oil royalties]. So, there's that. (SCO_GOV_02)

While tourism is often publicly visible in the capital city of Edinburgh, the oil sector has a significant but less noticeable presence. The two come into contact in specific spaces, for example in the Port of Leith or in the Firth of Forth, where flows of cruise ship or boat tour passengers encounter oil infrastructure. An oil sector participant defines the two as complimentary development paths as follows: "Tourism is hugely important for Scotland and Scotland is big tourist destination … I think broadly, the two seem to rub along quite well" (SCO_OIL_19).

However, there is little direct engagement through government-structured arenas of environmental governance. A government participant observes, "They're on a different scale and they don't tend to cross paths very often, [though] I think there are probably more links than you might perceive" (SCO_GOV_05). As in the Danish case, informal relationships serve as nascent connective arenas outside formal political arenas. This occurs in oil-as-tourism sites, which are spaces of educational outreach that craft narratives about oil development for visitors, rather than spaces for cross-sector dialogue that informs coastal planning or governance. Examples of these spaces include the Aberdeen Maritime Museum and the Aberdeen Science Centre, where the social importance of oil development is enacted for visitors and where the oil sector is largely defined as a positive economic force and a field of technological innovation. Another example is boat tours into the Firth of Forth, where tourists are informed that Hound Point terminal links ships and pipelines, creating connections between the oil hub city of Aberdeen and Edinburgh. The facility is run by BP, and visitors are told of the strong record of safety and environmental responsibility at the site. In these places, oil development is normalized as part of the tourist experience, presented as part of the seascape alongside nature-oriented tourism attractors like seals and seabirds, as well as historic attractors like Inchcolm island and abbey.

Conflictual Connective Arenas

The second type of connectivity also sees tourism and oil development as parallel development paths that typically do not come into contact. However, while both tourism and oil are seen as positive for host societies, there are also episodes of conflict.

In Newfoundland and Labrador there is an overall sense that offshore oil extraction and tourism development coexist as complementary modes of development. A tourism sector participant describes the mutual benefits of oil and tourism as follows:

> In regard to its impact on the tourism industry, it [the oil sector] creates a market, because we have a lot of people that are here on a contract basis. A lot of non-residents that are in for short periods of time, they're travelling. So, while they're here in the area, they're availing of a lot of the tourism activities as well, so they in turn become somewhat of a market for us. So, the overall general revenue that they've added to the provincial budget certainly goes into general revenue that comes back into

> the [tourism] marketing programs for the province. So, in that regard, it helps to diversify the economy as well. (NL_TOU_06)

Similarly, after noting that tourists rarely see operational oil infrastructure because it is "largely out of sight," an oil sector participant describes the complementarity of oil and tourism by saying, "The way I view it, because of the relatively parallel nature of the two industries, I see no reason why the oil industry wouldn't be supportive of the tourism industry. And I don't see why the tourism industry should be very concerned about the oil industry" (NL_OIL_04). While there is little evidence of government-structured processes or arenas that serve as connective arenas across the sectors, there are oil-oriented tourism spaces. The Johnson Geo Centre is an educational/tourism facility focused on the geology of Newfoundland and Labrador, which includes an oil-oriented exhibit that narrates the social importance of regional oil development for visitors. There are also nature-oriented education spaces that also serve as tourism sites, like the Manuels River Interpretation Centre, that receive oil sector financial support and sponsorship.

Exceptions to this coexistence emerge in conflict over proposed new oil exploration and development in places that have established tourism economies. This comes up in reference to proposed onshore-to-offshore hydraulic fracturing (or fracking) development near Gros Morne National Park, as well as proposed offshore oil exploration at the Old Harry site in the Gulf of St. Lawrence, off the west coast of Newfoundland (also see Smith 2016; Stoddart and Graham 2018; Stoddart et al. 2018). A tourism sector participant talks about their concerns with the Old Harry project by adopting the visitors' perspective:

> I come from an industrial area. And if I travel that far, I don't want to come and see another one. I want a pristine environment. I want the world as natural as possible. I don't want to take a picture of Gros Morne and say—oh, no, don't take a picture here, take a picture there because over there there's a flare burning. That kind of stuff. And they don't want to see that. They have money, they're willing to pay. But you have to give them what you want. (NL_TOU_18)

A government participant similarly describes the Gulf of St. Lawrence and western Newfoundland as a space where the two sectors cannot easily coexist because oil poses risks to established nature-oriented tourism economies. They say:

> The exploratory areas that need to come off the list first are all of those that pose a significant risk to the environment and other relatively more sustainable economies. So, the Gulf of St. Lawrence should be right off the list. It's a semi-enclosed sea, it's got counterclockwise circulation ... Oil platforms are a continuous source of small spills and they pose an unacceptable threat. ... When you have a beach and there are continuous oil slicks and tar balls and flaring on the horizon and dead birds and birds getting fried up... You know, our media will report all of that. And do you think tourists will really want to come and enjoy that? (laughs) I don't think so! (NL_GOV_31)

Environmental assessment forums and the Western NL hydraulic fracturing panel are identified as formal arenas for contesting oil development. These arenas include participation by oil and tourism players, as well as community groups and local environmental organizations. Where conflict between oil and tourism players emerges, the provincial government and C-NLOPB become mediators between the differing interests of tourism sector, oil sector, and civil society voices, who otherwise rarely come into direct contact. In these temporary connective arenas, tourism (along with local fisheries) is positioned as a more sustainable development pathway compared to the environmental risks of oil development. However, these temporary connective arenas are also critiqued by some participants as being overly aligned with oil sector interests and failing to provide fair spaces of deliberation and decision-making. The C-NLOPB is similarly characterized by a persistent failure to adequately regulate the industry and engaging in what an environmentalist participant calls the "farce of an environmental assessment" for proposed oil and gas proposals (NL_NGO_18).

The Norwegian case also has few connective arenas that incorporate oil sector and tourism players into environmental governance. While oil governance is nationally significant, tourism governance is more localized. A government participant describes the relationship between oil and tourism development by saying:

> I think compared to oil, you cannot compare it. Oil is much more important and we're using a lot more resources. The priority is low with tourism. But I think that it's because we have the oil. The economic side. If you see the revenue in tourist industries, we don't need tourism that much compared to other countries. (NWY_GOV_16)

Similarly, an environmental organization participant describes the division of oil and tourism governance by saying, "The tourism industry doesn't get nearly as much attention from the politicians as the oil companies do. I get the impression that in the minds of policymakers… tourism, that's more local politics. And that oil drilling is national politics" (NWY_NGO_02).

However, there are places where the coexistence of tourism and oil is more apparent. For example, the oil hub city of Stavanger sees regular flows of cruise ship traffic and boat tours to surrounding fjords, parks, and hiking trails. This is also the location of the Stavanger oil museum (Norsk Oljemuseum), the only educational and tourism space in our fieldwork devoted entirely to the history of the oil sector. This site presents a narrative of the oil sector as a valuable driver of economic development and technological innovation, as well as an important part of Norwegian history and culture. The museum also positions Norway as a global leader in oil and energy. At the same time, there are points of critical reflection on the oil industry, with displays of risk and disaster (the 1981 sinking of the Alexander Kielland platform), as well as an exhibition devoted to the intersections of oil, climate change and energy futures that juxtaposes perspectives from the oil sector and environmental groups, including the Bellona Foundation and Friends of the Earth. Similarly, Bergen is an important tourism hub, where tourism travel orients around UNESCO-designated historical sites, as well as access to the surrounding fjords, mountains, waterfalls, and hiking trails. Here, cruise ships and boat tours share space with oil infrastructure and travel routes. In these places, we see oil and tourism coexisting, but not integrated into connective arenas of environmental governance.

Conflictual connective arenas emerge in reaction to extending oil exploration and extraction into new areas (Dale 2016; Kristoffersen and Dale 2014). In our fieldwork, this most often comes up in relation to Lofoten Islands, where established fisheries and tourism economies are positioned as more sustainable then potential oil development. From this perspective, oil development poses risks to the natural basis of the tourism economy, including the fisheries that are integral to the heritage tourism identity of the region. For example, a government participant frames the risk of oil as follows: "If you have a blowout, for example. A possibility of that and also for the tourism reputation! Because Lofoten is nature-based and has clear waters that are unprotected and so on. So, what would an oil installation do to this?" (NWY_GOV_16). An environmental organization participant acknowledges that Lofoten is experiencing signs of negative environmental and social impacts of rapidly growing tourism in this remote rural region. However, they also point to the tourism sector as a more sustainable economic and social bulwark against extending oil development, saying:

> I think that if it wasn't for the tourism it would be much harder to protect this area against the oil industry because that is kind of… when you mention Lofoten for people, they already know the area, it's very famous. And they can't believe Norway wants to open this area for oil drilling because of the waste, nature, and everything. So, I think that overall, it [tourism] has a very good impact. (NWY_NGO_06)

Among community leaders that are more optimistic about the positive impacts of oil development, there is still a sense of caution that if oil development proceeds, it must respect the integrity of the established fisheries and tourism economies which are core to the identity and sustainability of the region (also see Kristoffersen and Dale 2014). When asked about oil development in the Lofoten region, a government participant responds with concern about the reputational damage for the Lofoten tourism sector, which relies on an image of romanticized unspoiled nature:

> Yes, would be bad for tourism. And also, Lofoten is a big local food area. And you think of Lofoten as a very clean nature area. So the food from the photos you think of it as pure and clean and fish in the land so with oil platforms in the distance you would get the same impression it would be bad for that. (NWY_GOV_17)

However, they follow this with the caveat that the two sectors could coexist under specific conditions: "But if you could ensure oil drilling that was someone looked at as responsible with the best possible technology used, I think that would be possible to work together with world heritage" (NWY_GOV_17).

Government-structured dialogue has created connective arenas, but there are challenges. A tourism sector participant notes, "Of course oil is much bigger in terms of GDP but even so you have big protection of companies in the oil industry and you have several thousand small travel businesses that don't necessarily cooperate" (NWY_TOU_19). While environmental groups have opposed oil development in Lofoten, the tourism sector is diffuse with varying opinions. While some local and regional tourism players have been vocal in their opposition, organizational voices for the Norwegian tourism sector have not asserted a strong opinion in opposition to oil development in the region.

Disconnected Development Paths

The third type of connectivity is disconnection, where one mode of development gets a lot of attention, while the other is marginalized from political debate and the social imaginary. Iceland is unique among our cases in that oil development remains hypothetical. In recent years, with global oil price declines and volatility, oil governance has moved to the margins of public and political discourse. An energy sector participant describes this as follows:

> Our research that has been done in the area (says) there may be oil in the area. There is a possibility there may be oil there. But nothing more than that. So, the hype of 2008-09 regarding Iceland becoming an oil nation… the hype was really weird, kind of. It was not based on very

> solid research at all. But there was this chance that there might be oil in a certain area in the north of Iceland, especially. The government decided to open up the continental shelf. And it has been a bit disappointing for the government and the people working on this. (ICE_OIL_07)

Further on in our conversation, the same participant summarizes the current situation by saying, "The oil prospects and the possibility of oil for Icelandic companies is totally a non-issue for Icelandic politicians now" (ICE_OIL_07). During the same period, the tourism boom evolved in counterpoint to the decline of oil development. The negative impacts of the tourism boom mean that needs for better tourism governance are higher political priorities. Attention is focused on issues like the rapid spread of AirBnB and loss of downtown rental housing, the unequal distribution of tourism benefits between the capital region of Reykjavik and much of the rest of the country, overcrowding of popular tourism spaces, as well as improving standards of conduct for visitors and tourism operators.

There is little evidence of formal connective arenas for environmental governance that integrates tourism and oil development. However, the reasons differ from the other cases. There are, however, notable themes at the margins of our data. While we are told that the tourism sector is not engaging with issues of oil prospectivity, a few participants observe that future oil development could challenge Iceland's self-image as a green society, which underpins its tourism destination image. For example, with regard to potential oil development, a government participant notes:

> Personally, I don't think that [oil] really goes hand in hand with a lot of what we're doing with the tourism industry here. But with municipalities that are struggling with survival, it might be a lucrative thought to be a part of it. But some big compromises would need to be done. And it just goes without saying that if you're trying build upon, for example, the whale watching business here, how long could you fit with the impact of industrial development here? And then the oil business? How can you manage that and make that harmony occur here in the harbour? It's a challenge. (ICE_GOV_22)

While not a central theme, this suggests the potential for conflict between tourism-oriented and oil-oriented visions for coastal development if oil exploration re-emerges as an issue.

Conclusion

This chapter examined the environmental governance dimensions of the oil-tourism interface. We examined horizontal governance processes involving a range of governmental and non-governmental players, as well as vertical governance processes involving different political scales from the local to the international (Francesch-Huidobro 2012). We extended this conceptualization of environmental governance to consider connective arenas where different sectors that share social-ecological space can engage in dialogue around planning, policymaking, and environmental management. Our concept of connective arenas echoes calls in environmental sciences and social sciences to think more holistically about landscape-level planning, as well as calls to shift from project-specific environmental assessments to more holistic studies of cumulative impacts of multiple forms of development (Berkes 2015; Schmidt et al. 2014; Young 2016).

We find some informal contact across the oil and tourism sectors. We also see that sites like maritime museums and science centres can work as cultural connective arenas for oil-oriented tourism and education. However, our main conclusion is that formal connective arenas of environmental governance are underdeveloped. Rather, connective arenas emerge during periods of conflict when new oil projects are proposed in places that have well-established and valued tourism economies, as is evident in the examples of Gros Morne National Park and Lofoten Islands.

The lack of connective arenas for engagement can be partly explained through understanding how horizontal and vertical governance processes unfold in these sectors. First, tourism governance involves a more diffuse network of players and orients around local and regional political arenas. By contrast, offshore oil governance involves a more concentrated network of players and orients around national and regional political arenas. Second, across most of our cases, offshore oil governance is better established than tourism governance, to the point where some scholars argue that environmental governance is overly influenced by oil sector interests in places like Canada (Davidson and Gismondi 2011; Graham 2019) and Norway (Dale 2016; Kristoffersen and Dale 2014). Third, while oil governance focuses on environmental assessment and social-economic benefits agreements, tourism governance focuses primarily on destination marketing and promotion. Attention turns to the negative social and environmental impacts of tourism in a reactive way after the problems of success become visible. This pattern is seen across tourism destinations that struggle with overtourism, such as Barcelona or Venice (Oklevik et al. 2019).

Our analysis has two major implications for broader questions about industrial and nature-based forms of coastal development. First, we argue for creating formal, ongoing connective arenas for engagement across nature-based and industrial forms of development, rather than treating connective arenas as responses to episodes of conflict. These connective arenas would allow for holistic dialogue about what forms of development are desirable and sustainable for coastal communities. This includes bridging the oil-tourism interface, but also connecting to fisheries and other vital sectors that are part of social-ecological networks. A significant challenge to implementing connective governance for coastal development is the multi-scalar characteristics of coastal and ocean environments, where the landward, shoreline, nearshore, and offshore zones fall within different political or managerial jurisdictions (Berkes 2015). Another challenge lies in navigating the dynamics of unequal political power and influence of different sectors. Cultivating connective forms of governance requires a political willingness to engage in experimentation and innovation, which may unsettle the political power of dominant players. Successful experiments and innovation can be "institutionalized" into new forms of governance (Gross 2010).

Second, our analysis reinforces other assessments that tourism governance and its relationship to sustainable development needs more attention (Gössling et al. 2012; Hall 2013; Snyman and Bricker 2019). For many regions, tourism governance is about destination promotion and building up tourist flows, with questions of managing social and environmental impacts emerging in a reactive way only after the negative impacts become visible. The current dominant approach to tourism governance is unlikely to balance the economic, social, and environmental wellbeing of host communities. Proactive approaches to envisioning and managing the types of tourism that are desirable for communities are more likely to meet the goals of sustainable tourism development. Furthermore, building stronger tourism governance could help level the political asymmetries between extractive and attractive development. This would help facilitate the creation of connective arenas for bringing these sectors into greater dialogue about development and social-ecological wellbeing in coastal societies.

References

Bach, M. (2019). The oil and gas sector: from climate laggard to climate leader? *Environmental Politics, 28*(1), 87–103.

Barnes, M., Bodin, Ö., Guerrero, A., McAllister, R., Alexander, S., & Robins, G. (2017). The social structural foundations of adaptation and transformation in social-ecological systems. *Ecology and Society, 22*(4), 16.

Berkes, F. (2015). *Coasts for people: Interdisciplinary approaches to coastal and marine resource management*. New York: Routledge.

Bryden, J. (2015). The development of industry and North Sea oil in Scotland and Norway. In J. Bryden, O. Brox, & L. Riddoch (Eds.), *Northern neighbours: Scotland and Norway since 1800* (pp. 123–153). Edinburgh, UK: Edinburgh University Press.

Bulkeley, H., & Betsill, M. (2005). Rethinking sustainable cities: Multilevel governance and the 'urban' politics of climate change. *Environmental Politics, 14*(1), 42–63.

Bulkeley, H., & Betsill, M. M. (2013). Revisiting the urban politics of climate change. *Environmental Politics, 22*(1), 136–154.

Castells, M. (2009). *Communication power*. Oxford, UK: Oxford University Press.

Ćetković, S., & Skjærseth, J. B. (2019). Creative and disruptive elements in Norway's climate policy mix: The small-state perspective. *Environmental Politics, 28*(6), 1039–1060.

Compas, E. (2012). "Retooling" for the new west: Environmental NGOs, planning, and governance regimes. *Society & Natural Resources, 25*(9), 883–899.

Cunningham, P. A., Huijbens, E. H., & Wearing, S. L. (2012). From whaling to whale watching: Examining sustainability and cultural rhetoric. *Journal of Sustainable Tourism, 20*(1), 143–161.

Dale, B. (2016). Governing resources, governing mentalities: Petroleum and the Norwegian integrated ecosystem-based management plan for the Barents and Lofoten seas in 2011. *The Extractive Industries and Society, 3*(1), 9–16.

Davidson, D. J., & Gismondi, M. (2011). *Challenging legitimacy at the precipice of energy calamity*. New York: Springer.

Di Gregorio, M., Fatorelli, L., Paavola, J., Locatelli, B., Pramova, E., Nurrochmat, D. R., et al. (2019). Multi-level governance and power in climate change policy networks. *Global Environmental Change, 54,* 64–77.

Dodd, S. (2012). *The Ocean Ranger: Remaking the promise of oil*. Halifax: Fernwood.

Dokis, C. A. (2015). *Where the rivers meet: Pipelines, participatory resource management, and Aboriginal-state relations in the Northwest Territories.* Vancouver, BC: UBC Press.

Duyvendak, J. W., & Jasper, J. M. (2015). *Breaking down the state: Protestors engaged*. Amsterdam: Amsterdam University Press.

Fischer, M., & Sciarini, P. (2015). Unpacking reputational power: Intended and unintended determinants of the assessment of actors' power. *Social Networks, 42,* 60–71.

Francesch-Huidobro, M. (2012). Institutional deficit and lack of legitimacy: The challenges of climate change governance in Hong Kong. *Environmental Politics, 21*(5), 791–810.

Gössling, S., Hall, C. M., Ekström, F., Engeset, A. B., & Aall, C. (2012). Transition management: A tool for implementing sustainable tourism scenarios? *Journal of Sustainable Tourism, 20*(6), 899–916.

Graham, N. (2019). Canadian fossil capitalism, corporate strategy, and post-carbon futures. *Canadian Review of Sociology, 56*(2), 224–250.

Gross, M. (2010). *Ignorance and surprise: Science, society, and ecological design.* Cambridge: The MIT Press.

Hall, C. M. (2013). Framing behavioural approaches to understanding and governing sustainable tourism consumption: Beyond neoliberalism, "nudging" and "green growth"? *Journal of Sustainable Tourism, 21*(7), 1091–1109.

Hilson, M., & Newby, A. G. (2015). The Nordic welfare model in Norway and Scotland. In J. Bryden, O. Brox, & L. Riddoch (Eds.), *Northern neighbours:*

Scotland and Norway since 1800 (pp. 211–229). Edinburgh, UK: Edinburgh University Press.
Kristoffersen, B., & Dale, B. (2014). Post petroleum security in Lofoten: How identity matters. *Arctic Review, 5*(2), 201–226.
Larsen, E. R. (2006). Escaping the resource curse and the Dutch disease? When and why Norway caught up with and forged ahead of its neighbors. *American Journal of Economics and Sociology, 65*(3), 605–640.
LeQuesne, T. (2019). Petro-hegemony and the matrix of resistance: What can Standing Rock's Water Protectors teach us about organizing for climate justice in the United States? *Environmental Sociology, 5*(2), 188–206.
Lövbrand, E., Hjerpe, M., & Linnér, B. (2017). Making climate governance global: How UN climate summitry comes to matter in a complex climate regime. *Environmental Politics, 26*(4), 580–599.
Mascarenhas, M., & Scarce, R. (2004). "The intention was good": Legitimacy, consensus-based decision making, and the case of forest planning in British Columbia, Canada. *Society & Natural Resources, 17*(1), 17–38.
McCauley, D. (2018). Reframing decommissioning as energy infrastructural investment: A comparative analysis of motivational frames in Scotland and Germany. *Energy Research & Social Science, 41*, 32–38.
Montpetit, E. (2003). *Misplaced distrust: Policy networks and the environment in France, the United States, and Canada.* Vancouver, BC: UBC Press.
Oklevik, O., Gössling, S., Hall, C. M., Jacobsen, J. K. S., Grøtte, I. P., & McCabe, S. (2019). Overtourism, optimisation, and destination performance indicators: A case study of activities in fjord Norway. *Journal of Sustainable Tourism, 27*(12), 1804–1824.
Parkins, J. R., & Davidson, D. J. (2008). Constructing the public sphere in compromised settings: Environmental governance in the Alberta forest sector. *Canadian Review of Sociology, 45*(2), 177–196.
Prior, T. L., & Heinamaki, L. (2017). The rights and role of Indigenous women in the climate change regime. *Arctic Review on Law and Politics, 8*, 193–221.
Rootes, C. (2013). From local conflict to national issue: When and how environmental campaigns succeed in transcending the local. *Environmental Politics, 22*(1), 95–114.
Sayles, J. S., Mancilla Garcia, M., Hamilton, M., Alexander, S. M., Baggio, J. A., Fischer, A. P., et al. (2019). Social-ecological network analysis for sustainability sciences: A systematic review and innovative research agenda for the future. *Environmental Research Letters, 14*(9), 1–18.
Schmidt, L., Gomes, C., Guerreiro, S., & O'Riordan, T. (2014). Are we all on the same boat? The challenge of adaptation facing Portuguese coastal

communities: risk perception, trust-building and genuine participation. *Land Use Policy, 38*, 355–365.

Shadian, J. M. (2014). *The politics of Arctic sovereignty: Oil, ice, and Inuit governance*. London: Routledge.

Smith, J. R. (2016). *When petro-capitalism comes knocking: Community interpretations and responses to the Gros Morne fracking controversy* (MA). Memorial University of Newfoundland, St. John's, NL.

Snyman, S., & Bricker, K. S. (2019). Living on the edge: Benefit-sharing from protected area tourism. *Journal of Sustainable Tourism, 27*(6), 705–709.

Stoddart, M. C., & Graham, P. (2018). Offshore oil, environmental movements, and the oil-tourism interface: The Old Harry conflict on Canada's east coast. *Sociological Inquiry, 88*(2), 274–296.

Stoddart, M. C., Smith, J. R., & Graham, P. (2018). Oil opposition. In D. J. Davidson & M. Gross (Eds.), *The Oxford handbook of energy and society* (pp. 447–460). New York: Oxford University Press.

Stoett, P. (2012). *Global ecopolitics: Crisis, governance, and justice*. Toronto: University of Toronto Press.

Young, O. R. (2016). The shifting landscape of Arctic politics: Implications for international cooperation. *The Polar Journal, 6*(2), 209–223.

5

Environmental Movement Conflict and Collaboration in the Oil-Tourism Interface

This chapter examines the role of environmental civil society players in relation to the oil-tourism interface. A focus on civil society players allows us to understand how connective arenas can be created independently from institutional political settings and governmental mediation. Oscillating between conflict and collaboration, interventions of civil society players in the oil-tourist interface suggest that innovation—and resistance to change—takes a range of patterns that are missed if we only focus on heightened periods of conflict and protest mobilization. As such, our analysis shows how public visibility and political efficacy, as aspects of social power, can be developed through social movement action in networks of conflict and collaboration. Civil society players are not just relevant forces in the present of the oil-tourist interface, but they also help envision and enact social-ecological futures for coastal societies. Building on marine sociology, this chapter highlights how the conflictual and collaborative strategic interactions used by social movement players helps shape public and political orientations to oceans as travel pathways, resource pools, or sites of environmental issues (Hannigan 2016; Widener 2018). We explore the emergence, existence, and maintenance of connective arenas across these sectors. At the same time, we examine

M. C. J. Stoddart et al., *Industrial Development and Eco-Tourisms*,
https://doi.org/10.1007/978-3-030-55944-1_5

the difficulties inherent to creating connective arenas, which leads to the parallel development of civil society interventions with the oil sector, on the one side, and the tourism sector, on the other side.

The chapter rests on the players and arenas approach and its conceptual toolkit (Jasper and Duyvendak 2015). This approach looks at social movements from a relational perspective and, in doing so, it locates social movement players in broader arenas of contention and collaboration. As James Jasper notes, this approach gives "equal and symmetric weight to protestors and to the other players whom they engage, and by focusing equally on players and the arenas in which they interact" (Jasper 2015, p. 9). This perspective highlights that civil society players engage with different types of arenas and other players. They develop strategic interactions with players that include corporations and other opponents, counter-movements, government agencies, the news media, and bystander publics (Amenta et al. 2017; Balsiger 2015; Han and Strolovitch 2015; McGarry et al. 2016).

Encounters among players take place in a variety of arenas, which include street protests or public demonstrations, courtroom trails, public input processes or consultations led by government or corporations, or media interviews. Jasper describes arenas as "a bundle of rules and resources that allow or encourage certain kinds of interactions to proceed, with something at stake. Players within an arena monitor each other's actions, although that capacity is not always equally distributed" (Jasper 2015, p. 14). The power of different players shift within different arenas, so that civil society actors that succeed at mobilizing large numbers of people in arenas of public protest, for example, may yet find themselves marginalized from the media or policymaking arenas. An application of this perspective is a re-examination of factionalization in the 1960s U.S. civil rights movement, which demonstrates how different movement factions made strategic use of the features of varied social, political, and media arenas (Polletta and Kretschmer 2015). Another example focuses on how the dynamic interplay of action and debate within the Brazilian student movement created new tensions and "subdivisions" within political arenas, as well as producing new relationships of collaboration and conflict (Mische 2015). This perspective also sheds light on how both the Occupy Movement and the Tea Party—despite dramatic ideological differences—engaged in similar processes of strategic alliance-building to grow these movements and gain visibility in public arenas in order to reach bystander publics (Han and Strolovitch 2015).

Starting from this analytical framework, this chapter tackles environmental movement interventions into the oil-tourism interface as follows. First, we look at the types of players that populate the arenas related to the oil-tourism interface and the relationships they have with other players in these arenas. Environmental groups may be involved in relationships with offshore oil and tourism organizations, as well as government players, that are both collaborative and conflictual. Looking across our cases, we see a general pattern of oil-government alignments and environmentalism-tourism alignments. While relationships with the oil sector are largely conflictual and focused on specific controversies over new oil exploration and extraction, there are also counterexamples of collaboration. This is especially in relation to projects devoted to renewable energy transitions and is more visible in the Norwegian case. By contrast, social network relationships with the tourism sector tend to be more collaborative, whether these are focused on aligning in opposition to controversial new oil projects or are focused on working to improve the environmental integrity of tourism practices. We find that Norway appears to have the richest environmental movement ecology among the cases, characterized by a mix of local, national, and international organizations that engage in collaboration and conflict across the oil-tourism interface.

Second, we examine types of strategic interaction that civil society players employ in different arenas, which we categorize as three different types. Patterns of punctuated conflict are the normal model of strategic interaction, where environmental movement intervention focuses on specific cases of new oil exploration or extraction. This normal model is characteristic of Newfoundland and Labrador, Scotland, and Denmark. The other types provide exceptions to this normal model. Patterns of conflictual collaboration are more evident in Norway. By contrast, patterns of collaboration oriented around tourism are more evident in Iceland.

While there are differences across specific episodes of contention, mobilization against new oil projects draws on collective action frames that include protecting wilderness and wildlife values, as well as protecting existing fisheries and tourism economies. In the case of Norway, we also see attempts to place regional oil controversies within the global context of climate change impacts and policy responses, including Norway's commitment to the Paris COP agreement. By contrast, mobilization in Newfoundland and Labrador contests the distribution of offshore oil benefits and risks within the region. In contrast to the oppositional collective action frames that are often used in relation to the oil sector, nature-based tourism is viewed in a more positive light as a sustainable development pathway and is often positioned as an alternative to offshore oil and extractive development.

When considering types of strategic interaction, we are also referring to a diverse range of arenas in which such interactions take place. Here, we see that civil society actors rarely engage in the normal, everyday operation of the oil sector. Instead, they tend to act in public protest or news media arenas. Mobilization occurs when new areas are proposed or opened for oil exploration and extraction, especially when this presents new risks to existing fisheries and tourism economies and social-ecological systems. This is evident in controversies over oil exploration in the St. Lawrence Gulf near Gros Morne National Park, Newfoundland and Labrador, and over extending oil exploration in the Lofoten Islands and Arctic in Norway (Stoddart and Graham 2018; Dale and Kristoffersen 2018; Kristoffersen and Dale 2014). By contrast with mobilization against oil development, civil society actors' engagement in the tourism side of the oil-tourism interface is more collaborative and employs less confrontational forms of action. Environmental movements engage in ecotourism or geotourism projects, or form coalitions in opposition to resource extraction. Environmentalist engagement in tourism is also driven by the need to ensure the environmental integrity of natural areas that serve as tourism attractors, an issue that is particularly visible during the Icelandic tourism boom.

The Players in the Arenas

When considering the types of players that engage in the tourist and oil sectors, there is a general pattern of oil-government alignment and environmentalism-tourism alignment. Despite this similarity, there are scalar differences in the arenas in which civil society actors play a role. In Denmark and Norway, international, national, and local players are all involved in the oil-tourism interface. Icelandic civil society players orient towards the national level, while civil society players in Scotland and Newfoundland and Labrador orient towards regional and local levels. There are, then, different configurations of players in the arenas that are more or less heterogeneous depending on territorial scope. More specifically, the five case studies display three different configurations of players. First, we have Denmark where international and national players seem to be at the centre stage, with some instances of grassroots mobilizations at the local level. Then, there is Norway, in which the configuration of players sees a balanced mix of international, national, and local players. Finally, there are Iceland, Scotland and Newfoundland and Labrador, in which national and regional players are more visible than international ones.

In Denmark, civil society players that come up repeatedly include the Danish Nature Conservation Association, Danish Ornithological Society, and Greenpeace. The former two are mostly active at the national level, while the third is a well-known international civil society organization that positions itself in international arenas as well. Although less central, WWF also has an internationally oriented role in the oil-tourism interface. Overall, then, a mix of national and international civil society actors seems to be dominant in Denmark. This is possibly a result of the COP15 meetings, held in Copenhagen in 2009, which were something of a tipping point. The aftermath of the 2009 COP meetings saw a decline in grassroots environmental activism. As a result, environmental movements currently work more through larger national and international organizations, which tend to be more professionalized and orient around policymakers more than around public or media debate. It should be also noted, though, that there are also episodes of localized contention around renewable energy and oil infrastructure that are driven by local communities, rather than larger institutionalized organizations that focus on coastal environments more broadly.

Norway provides an example of civil society players that are diverse regarding their territorial scope. There are national chapters of international organizations (Greenpeace, WWF, Naturvernforbundet—Friends of the Earth Norway), national civil society organizations (Bellona, Nature & Youth), and more localized, less professionalized groups (Oil Free Lofoten). This suggests the presence of arenas at different territorial levels with a mixture of civil society players ranging from more structured and institutionalized international NGOs to more grassroots local activist groups. This is consistent with the idea that the Norwegian government maintains an open relationship with NGOs. This configuration also helps explain the coexistence of a range of orientations to conflict and collaboration around oil and tourism development issues, from more moderate to more critical. A telling example is Bellona adopting a more moderate discourse of renewable energy transition that is more aligned with the framing of government and oil sector actors, while other ENGOs frame renewable energy transition through a more critical decarbonization lens (Lenferna 2018).

In Newfoundland and Labrador, ENGOs are less visible than in our other cases. There are chapters of national organizations, such as the Canadian Parks and Wilderness Society (CPAWS), provincial and local civil society organizations, as well as a couple broader interprovincial alliances that engage in specific issues: Save Our Seas and Shores and St. Lawrence Coalition. The key civil society players work primarily at the regional and local level, hence participating in arenas concerning the oil or tourist sectors from a more regional and local perspective. When looking at arenas related to specific controversies, we find a diverse range of players from the civil society sector, including: Sierra Club (Atlantic Canada chapter), Ecojustice, Council of Canadians, David Suzuki Foundation, Greenpeace, and WWF. It is mostly within localized arenas that we find a combination of international, national, regional, and local players that intervene in specific conflicts. Beyond these occasional exceptions, however, there are few central environmental civil society players and the field seems fragmented and relatively underdeveloped compared with our other case study regions.

In Iceland, there is a small core of national organizations and sporadic engagement with international groups. IceWhale, for example, is a civil society player that plays a role in national arenas and serves as a bridging tie between environmentalism and tourism. They focus on issues related to whales and whaling, while using tourism for public education and outreach. Other civil society players, like Icelandic Nature Conservation Society, are also positioned at the national level, while there is a more limited presence from international civil society actors like Greenpeace and WWF. Overall, key players are a small cluster of national organizations, with occasional involvement from international players. That said, the response to the 2008 crisis does come up as an exemplar of the tradition and potential for grassroots mobilization within Icelandic political culture, though this is less evident around the oil-tourism interface (Bernburg 2019; Hallgrimsdottir and Brunet-Jailly 2015).

In Scotland, a few core organizations come up repeatedly, with two players that seem to be most central. The Royal Society for the Protection of Birds is a Scottish/UK-based group that is more engaged in tourism-related issues, including tourism-ENGO collaboration around tourism as a site of environmental education. Greenpeace comes up in relation to conflict around oil and gas development. Scotland has a vibrant civil society at the local level (Diani 2015). However, a possible explanation for the absence of other national and international civil society players in the oil-tourism interface is that many larger ENGOs have their headquarters in London. Flows of activist capital and resources focus more on the capital and England, rather than on the regional Scottish level.

Two transnational ENGOs are central civil society actors in more than one case study region. These are WWF and Greenpeace, both as international ENGOs and through their national chapters. They have a role as international players and bridging ties between national/local politics and controversies.

Environmental movement players engage in strategic interaction in local, national, and transnational arenas regarding both the oil and the tourism sectors. While strategic interactions might be also looked at the micro-level of the individuals who participate in grassroots politics as activists (Jasper 2010), in our analysis we consider the meso-level of social movement organizations and other civil society players involved in various arenas. In other words, we look at strategic interactions among collective actors, rather than individuals, to illustrate both conflictual and collaborative types of strategic interaction within the oil-tourism interface and, when the connection between the two sectors does not happen, in both the oil and tourism sectors separately. First, we consider Newfoundland and Labrador, Scotland, and Denmark as cases of strategic interactions that revolved around patterns of punctuated conflicts, mostly around the energy sector. Second, we present the patterns of conflictual collaboration that are evident in the case of Norway, predominantly concerning the oil sector. Finally, we move to Iceland, which is a case of collaboration around tourism development.

Patterns of Punctuated Conflict in Newfoundland and Labrador, Scotland and Denmark

Patterns of punctuated conflict around specific cases of oil development expansion are the most typical form of strategic interaction. This form of strategic interaction is evident in Newfoundland and Labrador, Scotland, and Denmark. Here, we see the idea of tourism as a viable and sustainable alternative to the oil sector. Punctuated conflict also sometimes links local cases to the broader issue of climate change.

The oil-tourism interface is shaped through punctuated conflicts between ENGOs, oil companies, and government players. New oil developments are often contrasted with already existing tourism sites. ENGOs frame the former as risky endeavours that are not appropriate due to local ecological impacts or climate change implications, while the latter are depicted as sustainable enterprises that go together with the protection of local wildlife. There is a tendency, therefore, for ENGOs to align with tourism sector players.

In Newfoundland and Labrador, mobilization is more likely when proposed oil or onshore-to-offshore hydraulic fracturing projects risk impacting on the tourist sector at large and, more precisely, on specific tourism sites like Gros Morne National Park, or specific tourism activities like whale watching (Smith 2016; Stoddart and Graham 2018). In these cases, interventions focus on risks that oil brings to the environment for its own sake, but also to how it is valued within the tourism industry. An ENGO participant describes the social bases of mobilization around hydraulic fracturing as follows:

> People really felt like there was a level of risk when it came to their fresh-water supply, drinking water supply, with fracking. So, that really got people up in arms as well as the image of the oil and gas industry simply not going side by side with the image of the tourism sector, Gros Morne, a pristine area of the west coast. And perhaps there's a third leg—obviously, with the fracking, there was proposed an onshore-offshore in the coastal regions, which is obviously also the area where people go and make their livelihood from fishing. (NL_NGO_16)

ENGO players engage in sporadic conflictual interactions at the regional and local level, as happened in the case of proposed onshore-to-offshore fracking projects on the west coast of the island. Activists saw this a potential risk to Gros Morne National Park, which has positive economic benefits for the region. Overall, discourses that ENGOs develop in connection to their conflictual interactions are related to the wellbeing of Gros Morne and its natural habitat through a frame that presents oil and tourism economies as incompatible development paths. An ENGO participant describes this as follows:

> So, I think the thought of rigs, trucks, oil tanks, like all that is involved. I mean, when you look at the iconic images of Gros Morne, having that in the background, that would be hugely detrimental. You'd have to photoshop that out in all the tourist ads (laughs) and then people get there and be like, 'what is that?' ... And then the other biggest risk is spills. ... Even if you don't have a giant spill, there's chronic spills that damage your viability as a tourist destination. (NL_NGO_03)

Something similar happened in the case of proposed exploratory work at the Old Harry site in the Gulf of St. Lawrence, located offshore between the island of Newfoundland and the Magdalen Islands, Quebec. In these controversies, ENGO players develop collaborative interactions with tourism industry players, as well as with fishers, through coalitions like Save our Seas and Shores, or with the assistance of national groups like Council of Canadians. ENGOs consider oil risks not just to wildlife, but also to tourism and fisheries economies that are positioned as more sustainable alternatives to oil extraction.

The Fracking Awareness Network sums up these themes in a Facebook post, "fracking off the west coast of Newfoundland in the sensitive Gulf of St. Lawrence would be a terrible idea for our fisheries, communities, public health, tourism and coastal environments." The St. Lawrence Coalition takes these same discursive moves—keeping oil away from sensitive ecological areas—and applies it to the Gulf of St. Lawrence and locations like Shoal Point, near Gros Morne National Park. Another telling example of this is how CPAWS (Canadian Parks and Wilderness Society) intervenes using social media. For CPAWS, oil and gas development brings increased risks of oil spills, habitat destruction, and the loss of biodiversity. On Facebook, CPAWS fleshes out what is at stake by sharing detailed information about the species of birds, fish, and marine mammals that are threatened by potential oil spills that could impact the Avalon Peninsula on the east coast of the Newfoundland. This includes areas like Witless Bay, which is home to the Witless Bay Ecological Reserve and several boat tour operations oriented around whales and puffins. CPAWS asserts that government is allowing oil and gas exploration that puts these places at risk, even to the detriment of fish harvesters whose livelihoods would be threatened by an oil disaster. Nature NL does not tweet often but uses Twitter to focus on Witless Bay as a location at risk of spills and ecological damage from offshore oil.

The Old Harry controversy also demonstrates concerns around whales and ENGO-fishers alliances. Here, we see ENGO players having conflictual interactions in the policy arena with government players, including provincial ministries and the Government of Canada. In this case, ENGOs and other types of social movement players develop collaborative interactions with each other. Indeed, in this and other controversies that involve ENGO-government conflicts, several ENGOs collaborate to engage the public on opposing oil development projects through activities like letter writing to members of government. Similar collaborative interactions occur between ENGO players, academic players, and fishers. Such coalitions use a combination of strategic interactions in different arenas: legal actions in the court arena; calls for donations and volunteers in the public arena; letter writing and the development of trending hashtags such as #SaveOurGulf in the news media arena; and petitioning in the public protest arena. All together, these tactics lead to the development of multi-faceted conflictual strategic interactions with government players, while seeking support from the public. Through these tactics, ENGOs frame oil and gas employment as short-term and unsustainable in contrast to tourism employment as long-term and sustainable.

Environmentalist discourse that emerges in the media arena and through social media is about setting boundaries: keeping oil and gas companies out, while attempting to nurture more ecologically benign economic activities like nature-oriented tourism. The breaching of those boundaries, the moments when oil comes in, are framed through ecological destruction, habitat loss, and risk through oil spills or exploratory seismic blasting that may disrupt whales. The Sierra Club Atlantic Canada Chapter (ACC), for example use the #cdnpoli and #nlpoli hashtags to bring attention to oil and gas development and link local stories into larger discourses by using hashtags like: #oilandgas #saveyourgulf #oilspill #oilandgas #whales and #seismic. Sierra Club ACC calls out "greedy" oil and gas companies like BP, Energy East, and Irving by name, as well as the provincial and local governments that enable their behaviours.

The promotion of such discourse in the social media arena counterbalances action in other arenas where pro-oil players are hostile towards environmental protection discourses. Rather than adopting an explicit anti-environmentalist discourse, oil sector players focus on resisting overly onerous environmental regulations or assessment processes in the interests of remaining economically competitive in relation to other oil producing regions. At industry events, oil sector speakers rarely directly attack the environmental movement. However, invited keynote speakers who are not directly affiliated with the industry characterize environmentalists as "professionally motivated enemies and destroyers" of the oil industry, its vital economic impacts, and its contribution to "powering contemporary society" (field notes, June 20, 2017).

In the Newfoundland and Labrador case, ENGOs engage in sporadic conflictual interactions with government and oil industry players. When this happens, moreover, ENGO players rarely oppose the oil industry in general, but focus on the risks of oil development that impinges on established and highly valued tourism and fisheries economies. ENGO interactions in the environmental assessment and public engagement arenas are conflictual when it comes to interaction with oil industry players, but these arenas also allow for collaborative interactions with tourism industry players that become potential allies with whom to align. ENGOs, in short, more often engage in the oil-tourism interface by resistance to oil development, often in alliance with or on behalf of more sustainable tourism economies, rather than intervening around tourism development per se. Gros Morne National Park and other nature tourism sites are bases for opposing new oil development because of their established ecological and economic importance. As nature is re-commodified for tourism and revalued as the basis for a tourism economy, this provides leverage to ENGO opposition to new oil development.

In the United Kingdom, ENGO players like Greenpeace engage in public protest arenas around the issue of hydraulic fracturing (fracking), including advocating for the Scottish ban on fracking (https://www.greenpeace.org.uk/). A government participant describes the involvement of Greenpeace's and other ENGOs in the Scottish fracking debate as follows, "There were proposals recently to do fracking under the Firth of Forth and some of the old coal tunnels there and there were a lot of concerns raised there about the impact and whether the groundwater would be contaminated and whether it would cause structural instability and so forth. So, I think the Scottish government has sort of come out against fracking" (SCO_GOV_05). While there are historic examples of direct conflict between ENGOS and the oil industry, ongoing open conflict is more unusual. An oil sector participant describes the current relationship by saying, "NGOs like Greenpeace and Friends of the Earth obviously want Scotland to reduce its dependency on oil and gas. I think they also understand that you can't turn the taps off tomorrow. You need that transition" (SCO_OIL_19). However, in their written communication, Oil & Gas United Kingdom is critical of environmental movement players and characterizes Greenpeace in particular as "irresponsible" and "irrational" in its opposition to the oil and gas sector, while asserting the economic benefits of oil and that safety and risk mitigation is well done.

Greenpeace UK focuses on the risks of offshore oil exploration and drilling, though this is often focused not on the UK Continental Shelf, but on sites defined as globally significant, like the Arctic, Amazon reef, or Alberta oil sands. Through social media, ENGOs frame oil as a source of environmental risk as they also do in their official websites. Discussions of oil as a source of risk often link to seabirds and, more generally, the protection of biodiversity. However, this appears in the context of parallel discussions of other environmental issues including climate change risks and impacts, or biodiversity issues. In this regard, ENGOs on Twitter argue for setting boundaries that are impenetrable to the oil industry, with the linkages between oil development and climate change being particularly explicit. The discourse that Scottish ENGO players develop through their official websites revolves around bird and wildlife protection linked to whales, puffins, seabirds, seals, which also often appear as iconic animals and nature-oriented tourism attractors.

In Scotland, the discourses that emerge in the media arena through social media and ENGO websites focus on oil risks as part of a suite of environmental issues and concerns. Overall, there is not much recent mobilization or engagement around offshore oil issues. Rather, they tend to focus on anti-fracking campaigns, not just in connection to local arenas, but also to national and transnational ones. Scotland's WWF chapter articulates a discourse that links climate change and high emission levels to the need to not just keep oil and gas out of Scotland, but to inhibit it generally. They forward the notion that it is irresponsible and dangerous to increase oil and gas production, both offshore and through fracking, when current emission levels are simply too high. WWF Scotland uses Facebook to advance similar discourses, while also highlighting the risks that further oil exploitation poses to Scottish World Heritage sites, which is a small but notable link to tourism and its relevance for the Scottish economy.

Twitter, Facebook, and Instagram are digital media arenas where ENGOs circulate messages targeting oil sector players and, at the same time, render visible their direct actions from the public protest arena. Overall, the main targets are BP, Barclays and other extractive industry and related finance players. Greenpeace Edinburgh uses Twitter to create content that argues forcefully for an oil moratorium. On Facebook, Greenpeace opposes BP's movements to explore and drill more offshore oil wells in the North Sea. Hashtags like #BPshutdown, #climateemergency, and #nomoreoil, link the local case to global discourses about the climate and the behaviours of large energy corporations. In one Facebook post, Greenpeace frames BP as "desperately" trying to squeeze every drop of oil from the North Sea. Greenpeace employs Facebook to organize direct actions against BP—physically putting activists on oil rigs and in front of complicit companies like Barclays—that are then made visible through Twitter, where Greenpeace also stands in solidarity with other activists, like those at the gates of fracking facilities in Lancashire, England. Greenpeace also employs Instagram to spread its messages about oil development, through the savvy use of emojis and memes, and to further target oil industry actors like BP. For example, they overlay BP's logo onto photos and employs meme formats to convey their message—even using lines from Beyonce's "Single Ladies" to make a point about a BP oil rig and that we "can't afford to burn more oil."

The historical controversy over Shell's proposed sinking of the Brent Spar platform, in the Scottish North Sea, stands out as a critical moment that is still relevant for ENGO players in the region (Holzer 2010). The Greenpeace Brent Spar campaign has been memorialized as a particularly impactful period of contention. The Brent Spar controversy is raised at various points as a critical event shaping oil governance around decommissioning, including in the displays that shape oil-as-tourism narratives at the Aberdeen Maritime Museum. The museum creates a narrative that mobilization around this issue was a key political event for shaping oil governance regarding risk management and sustainability. This critical historical event provoked a re-think around issues of decommissioning within the oil sector and government. This shows how ENGO engagement in public protest arenas in the past became part of a shared memory about oil conflict in the region and had an impact on later oil-oriented policy measures. Similarly, an oil sector participant recalls this conflict by saying that in decommissioning:

> You need to do that in collaboration and agreement with other vested interests, like the fishing people as well. I don't think you should be doing it by yourselves. I think that is where, I suspect, Shell maybe got into a bit of trouble with [Greenpeace and] the Brent Spar. They didn't reach out to all the stakeholders, but they learnt their lesson. (SCO_OIL_02)

While observing that Greenpeace "told some porkies" (exaggerated) to gain media coverage, this participant concludes that the Greenpeace campaign was beneficial for provoking awareness of the "big picture" around governance and decommissioning that companies "have to get all the stakeholders involved and you have to find a way forward with that" (SCO_OIL_02). Conflictual interactions between Greenpeace and Shell over the Brent Spar platform are seen as a relevant historical case that is today part of the oil-as-tourism narrative at Aberdeen Maritime Museum. This conflict is presented as an important learning moment in a larger history of the social and economic importance of oil for Aberdeen, as a host community, and Scotland more broadly.

In Scotland, we also see examples of collaborative strategic interaction between ENGOs and players in the tourist sector. In Scotland, this happens regarding work on nature reserves and protected areas. Tourism-ENGO collaboration comes up with the Royal Society for the Protection of Birds (RSPB) as a key player, which is an ENGO but which also manages bird sanctuaries and nature reserves that are visited as part of wildlife tourism. The RSPB also provides awards that recognize "the inspiring people, projects and organizations working hard to protect Scotland's precious natural heritage," including notable tourism operators (field notes, March 23, 2017). There are examples of ENGO-tourism alignment and collaboration around tourism as a site of environmental education, with initiatives that revolve around activities like birdwatching aimed at increasing the environmental awareness of visitors. Exhibition displays at tourism sites also provide moments of public awareness of climate change impacts, as well as oil extraction as a source of risk for marine environments. Collaborative interactions are also linked with seabirds and wildlife protection. ENGO intervention in tourism develops around environmental education and the protection of local environments, with interactions occurring in local tourism arenas.

In Denmark, conflictual interactions between ENGO players and government players arise at the local level around renewable energy issues and participation in environmental assessment processes. In these conflicts, the main players are not formal institutionalized ENGOs, but rather rural community residents that gather as grassroots groups and engage in conflictual strategic interactions with government players around negative impacts of renewable energy, mostly focused on onshore windmills, which are seen as having potential negative impacts on coastal viewscapes. For example, a tourism sector participant describes oil and tourism as "very separate worlds, I think" but continues that there are some localized concerns that "you have to raise these high windmills offshore and they may affect the views from land and the views our tourists have from their accommodation sites and so on" (DEN_TOU_18). This is an interesting issue in terms of broader issues of climate change and renewable energy transitions.

During fieldwork for the project, there was also a localized conflict over a proposed heliport in the Wadden Sea National Park. The heliport would service offshore oil and wind power installations, including providing emergency and medical service with more rapid response times. The conflict is summarized by a government participant as follows:

> Tourism organizations, people who are in favour or nature or climate and so on say, a heliport in the middle of the World Heritage [site]? In the middle of the national park? On an island? Where people come for silence, quietness, calm environments? And then you have helicopters come all the time? No way. (DEN_GOV_08)

Pamphlets and signs about the issue were visible at information centres and other public places in the park, but this is a community-led initiative that is not affiliated with formal environmental or conservation organizations (field notes, April 26, 2017). For the most part, ENGOs did not voice major concerns or mobilize against the oil industry in Denmark.

However, some oil industry players see ENGOs as particularly conflictual towards them. An oil sector participant points to Greenpeace as a particularly difficult player to work with, describing their interactions as follows:

> We invited Greenpeace to come over. And they came over and handed over five copies of reports with the main point being that we should shut down the oil and gas industry completely. So, you know, we are pretty far away from each other. Because they do not acknowledge that we should have an oil and gas industry, so it's difficult. (DEN_OIL_08)

He contrasts this with the environmental think tank Concito, whom he sees as more reasonable and open to engagement because they have a shared "focus on CO_2 emissions" and a shared perspective that "these are complicated questions with complicated answers and not just one answer" (DEN_OIL_08).

The lack of major mobilizations led by institutionalized ENGOs is to some extent counter-balanced by conflictual interactions between grassroots groups and government and oil industry players at the local level on specific controversies, of which the No Heliport campaign in the Wadden Sea region is a telling instance. Where oil comes up as an object of contention for ENGOs beyond the local scale, it is around the expansion of Arctic oil frontiers. Conflictual interactions between ENGO players and oil industry players in campaigns against Arctic oil frontier expansion rely on frames of the Artic as globally significant, while linking it to climate change issues as well.

The lack of protest against the oil industry might reflect an ENGO focus on climate change as a separate and pressing issue. There is a lack of bridging between the two contentious issues, with climate change acquiring more centrality among environmental movements in the past few years. At the same time, though, this ENGO orientation may reflect the aftermath of the 2009 COP 15 meetings in Copenhagen, where massive protests were met with a strong, militarized police presence. Since then, ENGO action has largely shifted more towards policymakers and the political sphere, with grassroots and public or media-oriented activism less visible on the national scale. Similarly, during fieldwork the city of Copenhagen was in the process of "finding out about how we can withdraw from [fossil fuel] investments," though without the political provocation of a strong or highly organized social movement (DEN_GOV_08). When asked how the city adopted fossil fuel divestment, they explain:

> I think it was because Oslo actually started that. And again, with the whole green agenda and that we are promoting Copenhagen as a green and sustainable city. We are part of C40 ... It's an international organization for mega-cities, how to become more green and sustainable. And Copenhagen is obviously not a mega city. But we are a part of it because we are an innovator city. (DEN_GOV_08)

This institutionalization of environmental movement discourse and tactics is consistent with public projections of a Danish self-image of "global leadership in ... climate negotiations" and an image of "Denmark as a 'showroom' for green technology and economic growth" (Bjørst 2019, p. 130). This self-presentation of Copenhagen further complements its image as an urban eco-tourism site oriented around cycling and the harbour (Nilsson 2019).

Despite the lack of major mobilizations, two international ENGOs are particularly active in social media arenas across our cases. WWF and Greenpeace make significant discursive contributions to oil conversations in Denmark and in Northern Europe more broadly using social media like Twitter and Facebook. WWF is particularly active in countering the perceived need and inevitability of offshore oil development while folding in specifically Danish concerns, such as the need for public pension funds to divest from fossil fuel industries. At the same time, and especially on Facebook, WWF occasionally also addresses the tourist sector by urging cruise companies to be more responsible in their waste disposal and reporting the discovery of human waste on Danish beaches from cruise ships that operate in the nearshore zone. Greenpeace Dansk treads new discursive grounds by referencing the wealth of Scandinavia, hence contributing to the construction of a discursive space that goes well beyond Denmark and connects the Nordic region with the global scale of climate change: "A country like Norway, which has great wealth, also has a great responsibility to do more to solve the big, global crisis." Greenpeace also uses social media to highlight their direct actions, such as blocking a 27,000-tonne platform from the coast of Scotland. In this case, the international ENGO does not produce discourses that are limited to Denmark, but by speaking about a neighbouring region, they create a connection in the social media arena that goes beyond national boundaries when it comes to oil opposition. This goes hand in hand with the use of tweets to call out BP and their plans to erect more platforms in the North Sea as an ecological space that transcends national jurisdictions.

Within our fieldwork sites, collaborative interactions between Danish ENGOs players and tourism industry players seem rare and mostly involve nature-oriented NGOs in relation to Wadden Sea National Park. Tourism industry players and ENGO players' websites both promote the notion of tourism as a site of environmental education and sustainable development pathway. Overall, and despite examples of localized conflicts, in Denmark there is not a lot of direct engagement from ENGO players on either tourism or oil, but rather other concerns about coastal development and ensuring the health of aquatic environments. Key environmental issues that come up among ENGOs include agricultural impacts on coastal environments, bird and wildlife protection—which points to tourism as a space for environmental education—and overfishing. These themes are also linked to notions of the ecological value of coastal areas, nature as fragile and vulnerable, and the need to protect Danish coastal environments from human activities and overdevelopment.

Patterns of Conflictual Collaboration in Norway

Norway is a unique case when it comes to the type of strategic interactions that ENGOs players develop with other players in the public protest arena, the media arena, the policymaking arena, and the economic arena. Across these types of arenas, we see patterns of conflictual collaboration, where ENGOs engage in conflict and collaboration with government bodies and oil companies in different arenas around different issues (Kristensen et al. 2019). While some ENGOs are active in arenas for collaboration between the government and the oil industry, this does not prevent the emergence of a pattern of strategic interactions between ENGOs and other players that recalls Giugni and Passy's (1998) concept of "conflictual collaboration." In contrast to much social movement scholarship on institutionalization, the case of Norway is a prime example that the institutionalization of social movements into formal political arenas and collaborative relationships does not dull their critical capacities or make them apolitical (Masson 2015).

First, ENGO players are active in the public protest arena, where they pursue strategic interactions that involve moderate conflicts with oil sector players. This happens when the oil industry moves to expand oil frontiers into new territories. ENGO players engage in mobilization and contention in the public protest arena against oil industry players especially when projects for the expansion of oil exploration into new regions come into light, particularly Lofoten Islands (LoVeSe) and the Arctic. ENGO opposition is framed primarily around risks to fisheries and Arctic ecologies, but also around climate change and tourism impacts (Kristoffersen and Dale 2014). For example, an ENGO participant talks their groups' opposition to oil development as follows:

> That's why they [tourists] go to Lofoten, for the unique nature. … And I don't think that the oil platform is in coherence with this tourist image of Lofoten area. … One thing is that the infrastructure will probably be very ugly. But if there is an accident in this area it would be a disaster for the tourism industry, for the fishermen. For the cod that are born in this area, they're also travelling to other places in Europe and Russia. So, they would also suffer from that if there's a decline of the cod and other species. So that would be a disaster. The place where they have found oil is so close to the coastline, so there would be no time to stop. If there's a real accident, it would take a very short time for the oil to hit land along the coast. (NWY_NGO_06)

Protest messaging focuses on maintaining closed areas for oil exploration, climate change awareness, and claims about need for renewables transition and building sustainable employment in the renewable energy sector.

ENGOs engage in the public protest arena through online collective actions, which are not necessarily contentious, and collective actions that occur offline in the streets, sometimes involving more contentious forms of protest like rallies. Frequently, the public protest arena involves connecting online and offline spaces, which ENGOs exploit simultaneously to reinforce collective actions related to oil development. For instance, Greenpeace tweets in solidarity with activists climbing on the West Hercules rig offshore of Hammerfest, and thanks Greta Thunberg for her support of the People vs. Arctic Oil campaign that Greenpeace launched in partnership with Natur Og Ungdom (Nature and Youth). On YouTube, Greenpeace Norge gives visibility to direct actions related to the oil development issue. In one video, for instance, they interview activists on an Exxon Mobil rig outside of Olen, Norway. The goal is to stop Arctic drilling and the video also showcases the people making that happen. Another video shows an action that takes place during a fossil fuel conference, where activists dressed as polar bears line up in a conference room and are eventually arrested "we cannot sit here and see that this conference is moving forward because it is a bet against any responsible climate change policies."

While a general pattern of conflictual collaboration emerges in national arenas, interactions among social movement, government and oil industry players leans more towards conflict in specific local examples. For example, a September 2016 demonstration was held in front of the Stortinget (parliament) in Oslo. This was triggered by a government announcement of the latest round of bids for offshore oil exploration licences, which suggested the possibility of opening the LoVeSe region (field notes, September 9, 2016). ENGOs were confrontational towards government and oil industry players, as the choice to hold a public protest in the streets of the capital city illustrates. The coalition of groups participating at the protest included WWF, Greenpeace, Naturvernforbundet, and Nature and Youth. Political parties (particularly the Green

Party) and labour union representatives also participated. The decision to be active in the public protest arena helps keep the issue of climate change in public and political view. Several protesters also have large banners, smaller signs, or flags. Particularly visible protest slogans include, "Hands off the Arctic," "Save the Arctic," and "Climate justice, climate jobs, Greenpeace." For our ENGO research participants, nature-based tourism and renewable energy represent the kinds of "green jobs" that are positioned as desirable alternatives to fossil fuel extraction. There is also a banner from Nature and Youth specifically referencing the Lofoten region and articulating opposition to oil development there. Lofoten is also mentioned in speeches a few times as a region in need of protection against oil frontier expansion. A key part of the communicative strategy of the protest is to draw attention to Norway's "ecological irony," which Szerszynski (2007) characterizes as the gap between stated pro-environmental intentions and engaging in anti-environmental practices. Protest participants highlighted the irony of engaging in new oil exploration in closed areas—with special attention to Lofoten as a conflictual region—in light of Norway's professed commitments to climate change leadership. This also provides an example of how international ENGOs like WWF and Greenpeace magnify the concerns of national and local groups about the Lofoten region, thereby delocalizing the conflict and bringing it into the national political sphere.

Second, ENGOs engage in strategic interactions within the media arena using social media to reach attentive audiences and increase their awareness on oil-related issues, as well as to make requests for donations. An ENGO participant distinguishes their social media strategies as follows:

> In Norway, mostly only people who are politically involved somehow are on Twitter. Not regular people. So that would be more to lobby and discuss with oil industry. I've been arguing with Norwegian oil and gas on Twitter many times! And also, to approach politicians. And also, if you write on Twitter sometimes [mass] media pick up the comments. And they publish it. But for Facebook, I think is a very important platform for inspiring and mobilizing people in general [and reaching] a broader audience. (NWY_NGO_06)

Greenpeace Norge creates content on Twitter that highlights Norwegian cognitive dissonance when it comes to the climate and to the nation's continued exploration for oil in the Arctic: "the climate crisis is here, but Norway continues on autopilot as if nothing has happened." Hashtags like #savetheArctic link these discourses into larger conversations on Twitter. The same ENGOs also employs Twitter and Facebook to call out Equinor and financial institutions like Chase and JP Morgan, and the Norwegian state itself for supporting fossil fuel extraction in the Arctic. Other ENGOs, like WWF Norway, Naturvernforbundet (Friends of the Earth Norway), and Sea Legacy use Twitter to promote forms of collective action like political engagement, "awareness" campaigns, and online petitions.

ENGOs in Norway focus on the oil sector as a source of localized environmental risk. However, their concerns are also delocalized and scaled up to focus on risks and impacts of climate change, as well as government climate change performance and policy. In the Lofoten controversy the issue of envisioning post-oil societies also emerges as a recurrent theme. Anti-oil opposition in Lofoten is often framed in terms of fisheries impacts, as well as tourism impacts, though to a lesser extent. Sea Legacy's #OilFreeLofoten hashtag and campaign, which are very active on Facebook, highlight the ocean wildlife that is in the region, prefaced by the damage that oil exploration in Lofoten would do to these animals and their habitat. Sea Legacy also uses video to link oil and tourism in the Lofoten islands, citing the "huge sustainable fishery and tourism industry and massive cetacean populations." They show that for Sea Legacy, Lofoten is indicative of a pristine and specifically Norwegian idea of nature, one that is at risk due to oil development. This in line with what ENGOs participants talk about when thinking about future oil development in Norway. They advocate for moving towards a renewable energy social future, a discourse that is connected to oil's contribution to climate change, national policy responses to climate change, and opposition to extending oil exploration into Arctic oil frontiers. These social media communicative strategies work to delocalize the Lofoten controversy and make this region a symbol of national environmental concern.

In Norway, ENGOs develop discourses that draw attention to the ecological irony of a country that commits to fight climate change at the global level while continuing to invest in oil development projects at the national level (also see Røttereng 2018). On Instagram, for example, Greenpeace Norge and Naturvernforbundet both use images of oil rigs to bring home how literally close to nature offshore oil development is while framing oil development as an issue that transcends national borders. These ENGOs produce content clearly stating that oil development does not belong in Norway and cannot be expanded because climate change is a matter of global concern. ENGO discourse on social media is largely about disbelief that Norway would continue to exploit oil resources knowing full well the extent of global climate change. These ENGO players assert an uncompromising stance that new licences must not be granted and that oil reserves must stay in the ground.

Third, ENGOs are active in the policymaking arena, where they issue public statements or declarations that speak to the government and its main bodies. ENGO participants point out that there is a combination of conflict and collaboration with governmental institutions. An ENGO participant sums up this dynamic as follows:

> They [government] don't like when we make them look bad. But the principle of supporting NGOs, it runs deep. You can't remove the support for NGOs that we get from the public. … We are an open society. And it would look really bad, public opinion would crash down if they [government] — they've tried a couple of times, this government, to cut our support and the media storm always kept them at bay. (NWY_NGO_02)

Conflict revolves around the ecological risks of oil development and the notion that the government is dominated by oil interests. WWF Norway, Naturvernforbundet, and Sea Legacy all use Twitter to link oil exploration in LoVeSe to the ecological ruin it would bring to this "treasure trove" of Arctic nature and specific species such as whales, cod and other fish, as well as sea birds. In so doing, they make a clear connection between oil exploration and economic damage felt by fisheries and tourism sectors.

By contrast, collaboration between ENGOs and governmental players happens through strategic interactions with non-ruling political parties that can be effective in advancing social movement interests in the Norwegian multiparty political system. Furthermore, at the national level there are instances of collaborative strategic interactions across ENGOs and oil companies around renewable energy transition projects or risk mitigation. For example, when asked about engagement with the oil sector, an ENGO participant replies:

> Of course, we do. ... And, of course, one of the most important things there is to try to push them to be better in renewables, for instance, to invest more in the renewables sector. If you're only attacking them, then what's the point? Then, of course they won't listen to us. But if we talk to them as normal persons, as we do, then we can disagree but then we have the dialogue. (NWY_NGO_04)

A specific example of this type of mixed interactions is the relationship between social movement organizations and energy companies like Equinor (formerly Statoil) and the Norwegian Oil and Gas Associations, which activists discuss in terms of "critical engagement." Indeed, although ENGOs oppose them at times, some of them—Bellona and WWF for example—also collaborate or encourage work on shifting to renewable energy, or work on mitigation of the environmental risks of oil development.

The Lofoten controversy involves ENGO collaboration with some academic and opposition party players. It also involves sustained conflictual interactions with both government and oil industry players around the pressing issue of expanding oil frontiers and the broader theme of climate change. The presence of collaborative strategic interactions with some players and of conflictual strategic interactions with other players is also reflected in the tactic of producing public statements or declarations, as in the case of the Lofoten Declaration, which is a document that links a range of ENGO and academic signatories in opposition to the expansion of new oil frontiers (Lenferna 2018). The Lofoten Declaration invokes the Paris COP agreement to oppose further expansion of oil development and to hold government to stronger climate policy and performance. The Lofoten Declaration itself is a sort of "boundary object" that works to bridge the concerns of international, national, and local ENGOs and opponents of oil development and transforms this rural region of northern Norway into a symbol of global climate action against the further expansion of fossil fuel extraction (Slocum 2004; also see Lenferna 2018).

Fourth, when we shift focus to tourism in Norway, the economic arena seems particularly relevant for collaborative interactions between tourism players and ENGOs. This takes shape around specific projects, often linked to framing tourism as sustainable, sometimes in explicit opposition to the prospect of oil development as an unsustainable development pathway. There are signs of alignment between ENGO and tourism players around specific tourism sustainability projects, such as innovation to lower the environmental footprint of tourism, measures to improve corporate environmental responsibility, or projects around the electrification of vehicles, including boat travel. In some cases, ENGOs also critique the negative impacts of tourism in relation to cruise ships, tourism contributions to climate change, or impacts on Arctic nature. Overall, however, ENGO players support tourism as a sustainable alternative to oil development. When local risks of oil development impact on tourism, these are positioned as incompatible development paths with tourism seen as relatively more sustainable, but negatively impacted by new oil development.

The presence of conflictual collaboration across multiple protest, media, political, and economic arenas is in line with the commonly shared interpretation of Norway as an "open society," where there is a general principle of support for the rights of social movements to be heard within the institutional political arena (Kristensen et al. 2019). An ENGO participant describes this as follows: "You probably heard it before but in Norway it's quite open, there's easy access for us to talk to the government, there's easy access to talk to the political parties. … Bottom line, there's easy access and, of course, they don't listen to us every time but at least they are listening to what we have to say" (NWY_NGO_04). Furthermore, the range of tactics is consistent with our observations of Norway as a diverse social movement field that bridges larger-scale, more institutionalized movement groups and smaller-scale organizations.

Patterns of Collaboration in Iceland

In countries like Norway where the oil sector is well developed, ENGO players mostly engage in arenas related to oil. Something different happens in Iceland where ENGO players intervene in tourism sector arenas where there is collaborative interaction and alignment between ENGO players and tourism players. Among others, the group IceWhale is an important bridging tie that uses tourism as a space for environmental education, sustainability discourse, and anti-whaling campaigning.

ENGOs work through political arenas to preserve the integrity of natural areas and wildlife, such as whales, in ways that align them with tourism players. Conflictual interactions in political arenas have a strong connection with what happens in the economic arenas of the tourist sector. For example, environmentalist and tourism players see whales as a vital tourism attractor and part of tourism sustainability discourse (Cunningham et al. 2012). An environmental participant describes the environmental awareness value of tourism by saying:

> Operators that are running three-hour whale watching tours, they have an amazing platform to speak to people and educate them. They're doing this quite well and they're trying to get people to think about their impact and what they can do, just simple steps. … These nature tours are a good, inspiring opportunity and we have to make use to that. (ICE_NGO_07)

For example, The ENGO IceWhale is actively supported by many whale watching operators and has a focus on anti-whaling. The protection of whales as a tourist attractor reinforces collaborative interactions between ENGOs and one part of the tourist industry, which is focused on boat tours and whale watching, against another part of the tourist industry, which relies on Icelandic food culture and defines whale eating as part of tourism performance. In this case, collaboration between ENGO players and tourism industry players is far from homogenous, with a cleavage among tourism players around the conception of whales as objects of nature tourism versus whales as part of traditional food culture.

The campaign to extend Highlands National Park, which is supported by the Icelandic Nature Conservation Society and Landvernd (the Iceland Environmental Association), is another telling example of ENGO-tourism alignment and engagement with the state around ensuring protected areas against other uses, including hydro and geothermal energy infrastructure. At the same time, extending the park is a strategy for making it "easier to manage and control tourism impacts" (field notes, October 8, 2016). ENGOs and tourism players also share interests in improving recreational infrastructure for visitors in natural areas to ensure local ecological integrity in these tourism spaces. The protection and expansion of natural areas aligns ENGOs and tourism industry players in part because preserving wilderness means protecting the foundations of Iceland's tourism image and economy.

By contrast, the issue of oil development remains peripheral in Iceland. The connotation of oil as future-oriented and hypothetical, mostly focused on exploration activity, renders the oil industry rather marginal for ENGO players and their immediate environmental priorities. That said, when oil does emerge as an issue, then oppositional interactions emerge between ENGO players, oil industry players and government players. In our 2017 social media analysis, for example, we see the Pirate Party (opposition party) aligned with ENGOs in targeting government and opposing oil development, engaging in protest rallies against offshore oil and advocating for joining the Paris Agreement.

ENGO players also engaged in the media arena when oil development became a subject of political and public debate in Iceland prior to the 2014 price crash. An energy sector participant characterizes environmentalists' discourse in the media debate as follows: "Political debate about if Iceland should do this, if there is oil there why don't we just keep it in the ground. And even never explore it and give an example that we think this is not the right track for the world, bringing up more oil when we are, at the same time, fighting carbon emissions" (ICE_OIL_07). However, global oil price declines cooled the debate around future oil extraction in the country and ENGO players invested their resources in other campaigns. The same participant continues, "Well, over the last four or five years it's just off the agenda, really. It's not an issue. ... The oil prospects and the possibility of oil for Icelandic companies is totally a non-issue in Icelandic politicians now" (ICE_OIL_07). The lack of conflictual interactions around oil, then, reflects its peripherical position in political and public debate and not ENGOs disinterest in this issue. If oil re-emerges as a relevant issue in the political and public debate, ENGOs would be likely to re-energize opposition and campaigning.

Indeed, when it comes to energy issues, environmental movement players become active in political arenas against government and energy industry players in order to protect wilderness values. According to several ENGOs, natural areas need to be protected against industrial development, including hydropower plants and aluminium smelting that relies on Icelandic cheap, abundant energy. One of the issues that comes up recurrently related to renewable energy is that energy infrastructure and transportation lines negatively impacts wilderness values of Icelandic landscape, which in turn underpin the country's tourism image. Therefore, more than intervening in the oil-tourism interface, in the case of Iceland we are dealing with a broader "energy tourism interface."

Conclusion

Relationships between environmental and oil sector players are predominantly conflictual and focus on specific controversies. Contention is generally episodic, project specific, and triggered when new oil development impinges on established social-ecological relationships that include tourism and fisheries. This dynamic is most apparent in Norway and Newfoundland and Labrador. Where there is not an imminent threat of oil development impinging on already valued environments and economies, then mobilization or contention is less likely. Elsewhere, McAdam and Boudet (2012) examine a wide range of proposed energy projects in the United States and compare episodes of contention with the much more common occurrence of non-mobilization. They argue that social movements scholars' focus on episodes of contention gives a distorted view of the nature of social movement intervention in energy politics. Our research likewise shows that mobilization against oil development is more exceptional, rather than normal, and is related to the risks that new development projects pose to established economies and social-ecological relationships. In these cases, movement players increase their political efficacy by aligning with the interests of the tourism and fisheries sectors.

In our data from Norway, we not only see environmental movement conflict with the oil sector, but we also see signs of collaboration around renewable energy development. The Norway case is most characteristic of the logic of "conflictual collaboration" (Giugni and Passy 1998) between movements and governmental or energy sector players. Furthermore, in contrast to our other study regions, Norwegian political and public arenas are inhabited by a rich mix of social movement players that are rooted in local, national, and international political spheres. This ties back to notions of Norway as an "open society" with relatively good access for civil society players in policy arenas (Kristensen et al. 2019; Kristoffersen and Dale 2014). The shared notion of the open society is an important cultural resource for movement players to increase their access to—and influence within—political arenas.

The Scottish case also deserves closer attention here, as it highlights issues that are emerging around decommissioning oil infrastructure (McCauley 2018). The historical memorialization of the Brent Spar conflict, with Greenpeace as a key social movement player, is woven into narratives about present and future needs for safe and ecologically responsible decommissioning strategies. The memorialization of this conflict in spaces like the Aberdeen maritime museum, or in the personal recollections of our research participants, highlights the longer-term impacts that social movement players can make on cultural understandings of the oil sector and its responsibilities to coastal environments and host communities. The cultural impact of these conflicts also has political power effects as it leads to changing the standards and operating rules of political and economic arenas (Castells 2009).

By contrast, relationships between ENGOs and tourism show signs of alignment and points of collaboration, though there is also critical reflection the environmental costs and carbon intensity of tourism, particularly related to cruise ship travel and the ecological implications of expanding Arctic tourism. At times, environmental organizations and tourism sector players align around opposition to oil development in places with established tourism economies. In these cases, tourism is positioned as a sustainable development pathway in opposition to extractive development. However, environmentalism-tourism collaboration also involves specific projects for improving tourism sustainability practices, lowering carbon footprints, and environmental education. In the Icelandic case, we also see environmentalism-tourism alignment in opposition to whaling, wherein whales are valued as sources of tourist experience and representatives of Icelandic nature. These examples show that social movement players do not only exercise power through opposition and contention, but also through collaborative interaction and strategic alignment with other players. This is consistent with other research that suggests that nature-based tourism and environmentalism often have similar interests and rely on aligned discourses about the sustainability of attractive economies, even though working relationships across ENGOs and the tourism sector may be weakly developed (Cunningham et al. 2012; Stoddart and Graham 2018; Stoddart and Nezhadhossein 2016).

We also see instances of conflict around renewable energy issues, which shows that transitions away from fossil fuel economies may generate new points of tension and conflict (Burke and Stephens 2018). This highlights the importance of a broader focus on energy systems beyond oil in thinking about how environmental movements intervene in coastal development and energy futures.

The spaces of oil extraction and nature-based tourism that we examine are most often in rural settings and involve rural host communities. This raises several questions, including: Which players drive visions for rural development and shape the meaning of rural places? Are social movement interventions an example of rural development being re-imagined by urban civil society groups? In cases where conflict emerges over oil development, environmental movement action does not fit a clear rural-urban divide. Local and regional organizations with bases in rural communities are active in extractive resource controversies, but these conflicts also draw the interests of national and international organizations that play a significant role in (re)shaping visions for rural development across the North Atlantic.

In this chapter, we examined a range of ways in which environmental movement players intervene in multiple arenas related to the oil-tourism interface. This includes conflictual relationships and heightened periods of contention over oil development in political, public, and media arenas. It also includes collaborative relationships in the arenas of the everyday operation of the energy and tourism sectors. Through these varying roles, social movement players help shape political and public orientations to oceans, in terms of how they are valued as natural resource pools, travel pathways, or sites of ecological issues and political concern (Hannigan 2016; Widener 2018). Much social movement scholarship focuses on episodes of heightened contention. Going beyond our specific case studies, we show that the multiple roles that environmental movement players take in shaping extractive and nature-based forms of coastal development—beyond periods of heightened contention—deserve much more scholarly attention.

References

Amenta, E., Elliott, T. A., Shortt, N., Tierney, A. C., Türkoğlu, D., & Vann, B., Jr. (2017). From bias to coverage: What explains how news organizations treat social movements. *Sociology Compass, 11*(3), e12460.

Balsiger, P. (2015). Managing protest: The political action repertoires of corporations. In D. della Porta & M. Diani (Eds.), *The Oxford handbook of social movements* (pp. 653–666). Oxford, UK: Oxford University Press.

Bernburg, J. G. (2019). The cascading crisis and the changing base of popular protest: The case of Iceland. *Current Sociology, 67*(7), 1018–1038.

Bjørst, L. R. (2019). The right to 'sustainable development' and Greenland's lack of a climate policy. In U. P. Gad & J. Strandsbjerg (Eds.), *The politics of sustainability in the Arctic: Reconfiguring identity, space, and time* (pp. 121–135). London: Routledge.

Burke, M. J., & Stephens, J. C. (2018). Political power and renewable energy futures: A critical review. *Energy Research & Social Science, 35*, 78–93.

Castells, M. (2009). *Communication power*. Oxford, UK: Oxford University Press.

Cunningham, P. A., Huijbens, E. H., & Wearing, S. L. (2012). From whaling to whale watching: Examining sustainability and cultural rhetoric. *Journal of Sustainable Tourism, 20*(1), 143–161.

Dale, B., & Kristoffersen, B. (2018). Post-petroleum security in a changing Arctic: Narratives and trajectories towards viable futures. *Arctic Review on Law and Politics, 9,* 244–261.

Diani, M. (2015). *The cement of civil society: Studying networks in localities*. New York: Cambridge University Press.

Giugni, M. G., & Passy, F. (1998). Contentious politics in complex societies: New social movements between conflict and cooperation. In M. G. Giugni, D. McAdam, & C. Tilly (Eds.), *From contention to democracy* (pp. 81–108). Lanham, MD: Rowman & Littlefield.

Hallgrimsdottir, H. K., & Brunet-Jailly, E. (2015). Contentious politics, grassroots mobilization and the Icesave dispute: Why Iceland did not 'play nicely'. *Acta Sociologica, 58*(1), 79–93.

Han, H., & Strolovitch, D. Z. (2015). What the Tea Party and Occupy Wall Street illuminate about bystander publics as proto-players. In J. M. Jasper & J. W. Duyvendak (Eds.), *Players and arenas: The interactive dynamics of protest* (pp. 275–293). Amsterdam: Amsterdam University Press.

Hannigan, J. (2016). *The geopolitics of deep oceans*. Cambridge, UK: Polity Press.

Holzer, B. (2010). *Moralizing the corporation: Transnational activism and corporate accountability*. Cheltenham, UK: Edward Elgar.

Jasper, J. M. (2010). Social movement theory today: Toward a theory of action? *Sociology Compass, 4*(11), 965–976.

Jasper, J. M. (2015). Playing the game. In J. M. Jasper & J. W. Duyvendak (Eds.), *Players and Arenas: The Interactive Dynamics of Protest* (pp. 9–32). Amsterdam: Amsterdam University Press.

Jasper, J. M., & Duyvendak, J. W. (Eds.). (2015). *Players and arenas: The interactive dynamics of protest*. Amsterdam: Amsterdam University Press.

Kristensen, P. H., Moen, E., & Lilja, K. (2019). Civilizing business enterprises: The search for a new Nordic growth and development model. In S. Hanninen, K.-M. Lehtela, & P. Saikkonen (Eds.), *The Relational Nordic Welfare State: Between Utopia and Ideology* (pp. 202–223). Cheltenham, UK: Edward Elgar.

Kristoffersen, B., & Dale, B. (2014). Post petroleum security in Lofoten: How identity matters. *Arctic Review on Law and Politics, 5*(2), 201–226.

Lenferna, G. A. (2018). Can we equitably manage the end of the fossil fuel era? *Energy Research & Social Science, 35,* 217–223.

Masson, D. (2015). Institutionalization, state funding, and advocacy in the Quebec women's movement. In H. Ramos & K. Rodgers (Eds.), *Protest and politics: The promise of social movement societies* (pp. 79–97). Vancouver, BC: UBC Press.

McAdam, D., & Boudet, H. S. (2012). *Putting social movements in their place: Explaining opposition to energy projects in the United States, 2000–2005.* Cambridge, UK: Cambridge University Press.

McCauley, D. (2018). Reframing decommissioning as energy infrastructural investment: A comparative analysis of motivational frames in Scotland and Germany. *Energy Research & Social Science, 41,* 32–38.

McGarry, A., Davidson, R. J., Accornero, G., Jasper, J. M., & Duyvendak, J. W. (2016). Players and arenas: Strategic interactionism in social movements studies. *Social Movement Studies, 15*(6), 634–642.

Mische, A. (2015). Fractal arenas: Dilemmas of strategy and style in a Brazilian student congress. In J. M. Jasper & J. W. Duyvendak (Eds.), *Players and arenas: The interactive dynamics of protest* (pp. 55–78). Amsterdam: Amsterdam University Press.

Nilsson, J. H. (2019). Urban bicycle tourism: Path dependencies and innovation in greater Copenhagen. *Journal of Sustainable Tourism, 27*(11), 1648–1662.

Polletta, F., & Kretschmer, K. (2015). Movement factions: Players and processes. In J. M. Jasper & J. W. Duyvendak (Eds.), *Players and arenas: The interactive dynamics of protest* (pp. 35–54). Amsterdam: Amsterdam University Press.

Røttereng, J. S. (2018). When climate policy meets foreign policy: Pioneering and national interest in Norway's mitigation strategy. *Energy Research & Social Science, 39,* 216–225.

Slocum, R. (2004). Polar bears and energy-efficient lightbulbs: Strategies to bring climate change home. *Environment and Planning D: Society and Space, 22,* 413–438.

Smith, J. (2016). *When petro-capitalism comes knocking: Community interpretations and responses to the Gros Morne fracking controversy* (MA). Memorial University, St. John's, NL.

Stoddart, M. C. J., & Graham, P. (2018). Offshore oil, environmental movements, and the oil-tourism interface: The Old Harry conflict on Canada's east coast. *Sociological Inquiry, 88*(2), 274–296.

Stoddart, M. C. J., & Nezhadhossein, E. (2016). Is nature-oriented tourism a pro-environmental practice? Examining tourism-environmentalism alignments through discourse networks and intersectoral relationships. *The Sociological Quarterly, 57*(3), 544–568.

Szerszynski, B. (2007). The post-ecologist condition: Irony as symptom and cure. *Environmental Politics, 16*(2), 337–355.

Widener, P. (2018). Coastal people dispute offshore oil exploration: Toward a study of embedded seascapes, submersible knowledge, sacrifice, and marine justice. *Environmental Sociology, 4*(4), 405–418.

6

Lessons Learned and Social Futures: Building Social-Ecological Wellbeing in Coastal Communities

In this final chapter, we return to the overarching question raised in the Introduction: How can societies navigate the relationships between extractive development and attractive development in ways that contribute to social-ecological wellbeing for coastal societies and oceans? First, we elaborate the key similarities and differences across countries and regions of the North Atlantic to develop general insights for living with the oil-tourism interface. Second, we discuss the theoretical generalizability of our results beyond the oil-tourism interface and the North Atlantic regional context. We speak to how our analysis helps us better understand the complex networks of interactions among players involved in environmental issues and conflicts. We also speak to how our analysis advances our understanding of the ways in which culture, governance, and contentious politics are connected—or disconnected—in the environmental field. Third, and finally, we focus on lessons learned for building social futures in coastal societies. Here, we address tensions between different development pathways in a carbon-constrained world that requires low-carbon transitions. Climate change has been in the background throughout much of the book. We conclude by bringing it into the foreground, as the oil-tourism interface is also about climate

M. C. J. Stoddart et al., *Industrial Development and Eco-Tourisms*,
https://doi.org/10.1007/978-3-030-55944-1_6

change and the need for coastal societies to find development pathways that work in an increasingly carbon-constrained world.

Towards an Integrative Approach to Conceptualizing Environmental Issues and Conflicts

To synthesize our results, we return to the direct and indirect contact points that make up the oil-tourism interface:

- Energy as a tourism attractor.
- Conflict over energy exploration and extraction.
- The threat of energy extraction promotes tourism.
- Energy sector funding supports tourism development.
- Energy sector infrastructure supports tourism development.
- Tourism is fossil fuel and carbon intensive.
- Tourism environments are impacted by climate change.

Our results show signs of all these contact points to varying degrees, except for the notion that the threat of energy extraction promotes tourism. In other research, sociologist Patricia Widener (2011) found that the threat of oil development increased tourism flows to parts of Ecuador that were in the path of proposed pipelines. Tourism flows were shaped by the desire to see these landscapes before they were disrupted by imminent oil development. However, even in cases of controversies and conflict over oil extraction, this does not emerge as a notable theme in our analysis.

Conflict Over Energy Exploration and Extraction

Looking at direct contact points, we see examples of conflict over oil exploration and extraction. Across cases, these conflicts are episodic and exceptional, rather than an ongoing feature of the eco-politics of oil and tourism development. The ongoing mundane operation of oil extraction across the North Atlantic proceeds without a great deal of opposition. Across our case studies, the oil sector is viewed positively in terms of their social and economic contributions to host societies, while governance regimes are largely viewed as ensuring safe and environmentally responsible resource extraction. Rather, conflict emerges in specific cases where new oil exploration or extraction impinges on established, valued tourism and fisheries landscapes and seascapes. As noted in Chapter 1, there are three key facets of social-ecological relationships with oceans as travel pathways, natural resource pools, and sites of ecological problems and political contestation (Hannigan 2016; Longo and Clark 2016; Urry 2014; Widener 2018). Based on our research, conflict is more likely when different orientations to oceans as natural resource pools and as pathways for tourism and leisure travel come into direct contact. In these cases, social movement players align with tourism interests to help increase their political efficacy against the power of the oil sector. This is most apparent in controversies over oil development in the Lofoten Islands region of Norway, or around onshore-to-offshore hydraulic fracturing and offshore oil exploration in the Gulf of St. Lawrence near Gros Morne National Park in western Newfoundland (also see Dale 2016; Dale and Kristoffersen 2018; Kristoffersen and Dale 2014; Smith 2016). Others have documented friction in parks governance between protecting ecological integrity and economic development models of maximizing visitation (Campbell 2019). However, in our study, we also see how parks and protected areas, such as Gros Morne, become important leverage points for opposition to oil development. Nature gains political standing and voice from environmental groups and tourism sector interests because it can be translated into economic value for tourism host communities. This is consistent with other research on tourism-oriented opposition to oil in sites such as Belize and Ecuador (Gould 2017; Widener 2011). We also see examples of opposition to Arctic oil development in Norway and Denmark. However, this is less based on concerns about impacts on tourism economies, but is rooted in concerns about ecological values and the discourse of a global Arctic: an Arctic that is represented as the object of global environmental change, scientific inquiry, and political concern (Bravo 2009; Stoddart et al. 2019).

Energy as a Tourism Attractor

Another direct point of contact is that oil can also serve as a tourism attractor. This is not prevalent across our cases. However, most sites do have examples of tourism and educational sites such as museums or science centres that include exhibitions and displays about the oil sector. These sites, such as the Esbjerg Maritime Museum (Denmark), Geocentre (Newfoundland and Labrador), Norwegian Petroleum Museum (Norway), and Aberdeen Maritime Museum (Scotland) often enact narratives about oil and its social and economic importance for host regions. For the most part, the oil sector is presented in a positive light for visitors, with few examples of reflexivity about the limitations or social-ecological challenges of oil (Davidson and Stedman 2018). The most notable exception is arguably the Norwegian Petroleum Museum, which provides space to explore the climate change implications of oil and for the presentation of environmental movement voices. Scottish sites, including museums and science centres, similarly bring in the notion of the "energy trilemma," or the need to balance economic sustainability for oil host communities, growing global human development and energy needs, and the imperative of responding to climate change. In the Scottish case, we also see how historical episodes of conflict over offshore oil—particularly the Brent Spar conflict—have become memorialized as a critical event that is shaping contemporary oil sector practices around sustainability and decommissioning. These examples provide more nuanced and complicated discussions of the role of oil, including occasionally offering space for dissenting voices. Museums and science centres serve as physical spaces where visitors encounter narratives about oil and its impacts on host regions. They offer a glimpse into the narratives that oil host communities create and promote as part of creating a regional identity.

Energy Sector Funding Supports Tourism Development

In his writing on Shetland Islands, Andrew Jennings (2015) notes that oil development was leveraged to support nascent tourism development. Though this is marginal in our data, we see some examples of flows of resources from oil companies to support tourism development. Most often, this takes the form of funding and corporate sponsorships to sites like museums or science centres where there are specific oil-oriented exhibits. However, we also see supports for other cultural events and facilities, as well as sponsorships for outdoor recreation infrastructure like hiking trails. These are not solely tourism oriented. Rather, they serve multiple roles as local community amenities that benefit from oil sponsorship, as well as being integrated into the tourism infrastructure and visitor experience of these places. However, these contact points have the potential to become contentious. For example, social movement players have raised provocative questions about the role of oil sector sponsorship for arts institutions like the Tate art museums (United Kingdom) because of climate change concerns (Motion 2019).

Energy Sector Infrastructure Supports Tourism Development

Direct contact points between oil and tourism are infrequent, but show how the sectors come together through conflict, collaboration, or flows of resources. Shifting our attention to indirect contact points, one idea that comes up occasionally is that infrastructure to support the oil sector also benefits tourism development. This theme comes up in discussions in Aberdeen (Scotland) and Esbjerg (Denmark) as evidence for the positive spillover effects of oil development. Greater flight connectivity, as well as hotel and restaurant development, serve the needs of both oil development and tourism travel. For several participants, this supports the idea that oil and tourism can work as complementary development paths.

Tourism Is Fossil Fuel and Carbon Intensive

Another indirect contact point is that tourism itself is carbon intensive. Flows of oil ensure the multiple modes of travel that provide tourism mobility, including airplane, car, and bus travel, as well as the various aquamobilities of cruise ships, ferries, and boat tours. These carbon-intensive mobilities are deeply woven into tourism and come up repeatedly across cases. However, this is generally not problematized, but is taken for granted as part of the background infrastructure of tourism travel and experience. The carbon intensity of tourism mobilities is a particularly difficult problem in achieving the objectives of sustainable tourism (Cohen et al. 2011; Peeters et al. 2019; Young et al. 2015). However, across cases there are few signs of critical engagement with this issue. Environmental reflexivity about the carbon intensity of tourism travel is more evident in our interviews than in other data sources. The Icelandic case appears to show more reflexivity about the carbon cost of tourism. This may reflect the experience of the rapid tourism boom in the region, coupled with Iceland's status as a remote destination that relies on airplane access for virtually all of its tourism market (Lund et al. 2017). Here, tourism is more often viewed as relatively sustainable, but the carbon footprint from plane and car travel is an issue to be addressed. Across several of our cases, participants and website data point to specific projects that are emerging that attempt to reduce the carbon intensity of tourism by experimenting with the electrification of car, bus, or boat travel. These localized experiments have the potential to be scaled up and become positive tipping points in shifting towards lower-carbon tourism mobilities.

Tourism Environments Are Impacted by Climate Change

The final contact point is that high northern tourism environments are impacted by climate change, which in turn is driven by fossil fuel extraction and consumption. Where climate change is discussed in our data, this is usually broader than impacts for the tourism sector.

Climate change concerns orient around coastal erosion, sea level rise, flooding, melting glaciers, and extreme weather. There is also concern that these ecological and landscape changes are amplified in the high north (Callison 2014; Cunsolo and Ellis 2018). Climate change has spill over effects on tourism economies by altering coastal landscapes through coastal erosion or destruction of infrastructure (Hall et al. 2009). This poses challenges to tourism development in general, but also has specific impacts on practices like hiking and boat tours, which are key tourism activities throughout the region. Shifting seasonal weather patterns can also impact tourism, for example by posing challenges for winter activities and attempts to extend the tourism season. Conversely, there is also material in our data about how climate change is also opening new avenues of tourism activity in the high northern reaches of the North Atlantic. With less sea ice and longer ice-free seasons in the Arctic, there is increasing cruise ship travel, as well as increasing interest in Arctic tourism that draws visitors that want to experience these places before they are irrevocably transformed by climate change (Hall 2010; James et al. 2020). The notion that climate change will re-shape possibilities for tourism in northern oceans demonstrates the interactive and evolving nature of the relationship between oceans as travel routes (e.g. Arctic cruise tourism) and oceans as sites of ecological issues (e.g. climate change and melting sea ice).

Living with These Contact Points

In terms of living with offshore oil, we argue that Norway and Denmark serve as positive models for oil sector host communities in a climate changing world. Both Norway and Denmark use government intervention to maximize the social-economic benefits of oil for host communities, as well as to support wind power development. In both cases, we see more evidence of environmental reflexivity around the tensions between oil development and living in an increasingly carbon-constrained world, which necessitates renewable energy transitions (Davidson and Stedman 2018). It is striking that these regions offer similarly positive models, given their differences in their political economy orientations and evaluative repertoires. Norway is relatively high-oil, low-tourism in its political economy orientation, while the industrial order of worth is more prevalent in our data. By contrast, Denmark is relatively low-oil, high-tourism in its political economy orientation, while the ecological order of worth is more prevalent in our data. At the same time, both countries are positioned at the consensus-oriented and corporatist end of the political spectrum (Lijphart 2012). Both countries are also characterized by coordinated market economies (Hall and Soskice 2001). These deeper political and economic characteristics likely help explain the similarities we see around managing oil development, planning for renewable energy transitions, and environmental reflexivity.

By contrast, the Scottish case highlights issues around decommissioning oil infrastructure in post-peak fields (McCauley 2018). This emphasizes the need to take a long view of oil development and to plan for how host communities can successfully transition away from oil sector dependency. This echoes energy justice research, which argues that planning for energy transitions needs to be attentive to distributional issues related to the risks and impacts of energy development, as well as procedural issues related to the meaningful engagement of host communities in decision-making and planning (Jenkins et al. 2016; Le Billon and Kristoffersen 2019; Sovacool et al. 2017).

Environmental reflexivity is often provoked by social movement players that challenge the expansion of oil development. Here, we highlight the constellation of Norwegian social movement players as an important factor. Norway is characterized by social movement players that include national chapters of international organizations, Norwegian national organizations, and local grassroots groups, which operate in arenas of an open political culture. Movement players draw attention to the Norwegian paradox of expanding oil development while claiming to work for decarbonization. For example, Greenpeace Nordic and Nature and Youth have used the courts as an arena to challenge the Norwegian government over oil exploration and extraction in the Barents Sea, which translates into greater public attention to these issues.

The Icelandic case, which like Denmark is more oriented towards an ecological order of worth, shows that oil prospectivity can move in and out of public and political visibility. Interest in Arctic oil exploration waxes and wanes in response to global oil prices and current global price volatility makes the region less desirable. We argue that it is important for policymakers, social movements, and civil society to engage in discussions of how offshore oil development may (or may not) fit into visions for social-ecological futures, rather than waiting for a rebound in the economics of Arctic oil exploration before this re-emerges as a subject of public and political discussion.

Turning to nature-based tourism, across much of the region tourism governance is not viewed as a political priority but emerges in a reactive way once the social and environmental challenges of success become apparent. While a great deal of resources goes into promoting tourism destinations and attempting to increase visitor flows, less attention is given to how governance can help ensure *sustainable* tourism for host communities. Particularly in the Icelandic case, but also in Lofoten Islands (Norway), tourism governance did not arise as a priority until the negative impacts already became apparent. As parts of Norway are experiencing the tensions of overtourism, one response is to shift tourism promotion to the idea of "all of Norway, all year round" to diffuse tourism beyond the most popular hubs (VisitNorway 2019). Recent developments in Iceland also raise questions about the economic and social sustainability of the tourism sector (Shaban 2019; Statistics Iceland 2019). The tourism boom has softened in the wake of the 2019 bankruptcy of the budget WOW airline and in response to the rising Icelandic Króna and cost of living. The drop in tourist numbers raises concerns that the tourism sector may not be economically sustainable in the long term but may also entail challenging boom and bust cycles. We argue that tourism governance should be developed in parallel with tourism promotion and marketing to ensure that tourism development is managed for the social-ecological wellbeing of host communities. This approach to tourism governance should also incorporate the meaningful participation and expertise of civil society and social movement players, who often promote tourism as a sustainable development pathway and alternative to resource extraction. This will help ensure that tourism governance is used to proactively structure tourism development to better meet the environmental, economic, and social needs of host communities. Among our cases, Scotland is closest to this model, with a good flow of dialogue between tourism sector organizations and government, which connects tourism promotion with policymaking and visioning how the tourism sector should develop to meet economic, social, and environmental sustainability goals. By broadening the perspective on tourism to include economic, social, and environmental sustainability goals, Scotland also goes beyond common notions of ecological and inspirational worth to "value" tourism in terms of market and civic worth. This opens tourism governance to a broader range of players.

Living with the contact points between oil and tourism development involves other possible tensions around economic sustainability. For example, if oil development increases the cost of living in a host society, as in Norway, does this shape the extent or type of tourism development that is possible? Could oil development unintentionally crowd out tourism development, or limit it to more affluent visitors? Similarly, natural resource economies are subject to boom and bust cycles. Tourism flows are also shaped by global trends and current events, such as the 2020 COVID-19 global pandemic. Boom cycles in tourism or natural resource development may suddenly increase costs of living, which may create anxiety or resistance within host communities. Although these are not strong themes in our data, these questions deserve attention in future research, as we consider the importance of economic sustainability in building social-ecological wellbeing for coastal societies.

Offshore oil and nature-based tourism are often viewed as parallel development paths, which demonstrates how oceans can be simultaneously valued as natural resource pools and as pathways for tourism and leisure travel. While oil is primarily valued in terms of economic wellbeing (within the industrial and market orders of worth), tourism is valued across multiple orders of worth. Tourism benefits are seen as more multi-dimensional, contributing to economic, social, and cultural wellbeing, as well as contributing to rural communities more than oil development. This is consistent with our previous research on tourism development in Newfoundland and Labrador, which finds that the perceived benefits are social, cultural, and environmental, as well as economic (Ramos et al. 2016). There is generally a broad, cross-sectoral consensus about the positive impacts of tourism as a relatively sustainable development pathway. By contrast, interpretations of oil are more heterogenous and sometimes conflictual. Where there is conflict around oil development, this often breaks into oil-government alignments versus environmentalist-tourism alignments (also see Stoddart and Graham 2018). While these sectors may share social and ecological space, they rarely share political and cultural space.

Contact points across the sectors become most visible when there is conflict over proposed new oil exploration and extraction, which happens when orientations to oceans as natural resource pools directly come up against orientations to oceans as pathways for tourism and leisure travel. Contact points also emerge where oil itself becomes a tourism attraction, or when oil sector funding helps support tourism development. Building on our arguments about evaluative repertoires and orders of worth from Chapter 3, we argue that there is space for much greater engagement across the sectors. Offshore oil and nature-based tourism represent substantially different pathways for how we live with, and make a living from, coastal areas. However, planning and policymaking for these sectors often occurs in silos rather than through engagement across sectors. Chapter 3 demonstrates that news media discussions about oil and tourism rarely intersect. Public deliberation is characterized by "cultural holes" in our collective understanding that obscure the direct and indirect contact points between these two development models. In other words, the oil-tourism interface is a cultural blind spot across our five case study regions. Furthermore, these sectors are further impacted by, and contribute to, climate change, which is also treated as a separate policy arena. As we argue in Chapter 4, there should be more regular contact points for communication across sectors that share social-ecological space, so that their divergent interests are not only visible during times of conflict. These contact points should also provide space for meaningful participation by civil society and social movement players, as noted in Chapter 5. This would help build more holistic landscape-level and seascape-level forms of environmental governance, which would contribute to greater social-ecological wellbeing for coastal communities and oceans.

Relational Political Ecology: Connecting Environmental Governance, Players and Arenas, and Culture in Action

We approached this project from a theoretical perspective we call relational political ecology. While there are multiple ways to define political ecology, there is a common emphasis on understanding the interconnections between ecological, political, economic, and cultural systems for structuring social-ecological relationships across the mirco-, meso-, and macro-level of networks and processes (Bennett 2010; Escobar 1999; Latour 2004; Stoddart 2012). This perspective attends to multiple facets of social power, including political economic significance (Chapter 2), cultural visibility (Chapter 3), and access to—and efficacy within—public and political arenas of deliberation and decision-making (Chapters 4 and 5). By framing our approach as relational political ecology, we align with other complementary relational sociologies that include cultural and social network analysis, a comparative analysis of evaluative repertoires, and the players and arenas approach to social movements (Crossley 2010; Duyvendak and Jasper 2015; Jasper and Duyvendak 2015). We integrate these approaches because they provide specific conceptual tools that help us analyse culture, governance, and social movements within a relational political ecology perspective. By drawing these theoretical threads together into a broader relational political ecology framework, we emphasize the social-ecological contexts and relationships that culture, governance, and social movement players work through and within. As such, our relational political ecology approach offers a valuable theoretical perspective for future studies of how societies navigate, negotiate, or contest natural resource extraction and ecotourisms as alternative development pathways.

In the cultural dimension, both tourism and oil are valued for making economic and social contributions to host communities; in other words, oceans are valued within multiple "orders of worth" as natural resource pools and travel pathways. However, in the political sphere there is often an asymmetry that privileges oil over tourism, as in Norway or Newfoundland and Labrador, which are defined in part by their general orientation towards industrial and market orders of worth; or where tourism is more visible than discussions around oil, as in Iceland or Denmark, which are oriented more towards ecological and market orders of worth. Even where there is more symmetry across the oil and tourism sectors, as in Scotland, the connected dimensions of governance are underdeveloped and the two sectors are treated as separate policy arenas. The connective dimension of environmental governance emerges only episodically and reactively when conflict erupts, driven by the opposition of community members or environmental movement players. By thinking of the relationship between resource extraction and tourism development through a relational political ecology lens, one of the main arguments we make is to call for more connected forms of environmental governance in order to encourage planning for relationships among multiple development paths at the landscape/seascape level. This call is in line with other strains of thinking in environmental sciences around the need for holistic landscape or seascape level environmental impact assessment, or cumulative effects analysis (i.e. Berkes 2015). A more holistic approach would resolve some of the tensions that result from the dominant disconnected approaches to environmental governance. One of the challenges to building more connective forms of governance involves navigating the economic, cultural, and political power differentials across sectors and actors. However, more holistic forms of environmental governance could provoke significant shifts in how we envision the social-ecological futures that host communities want and who has the political efficacy to enact these futures. As such, our concept of the connective dimension of environmental governance makes a valuable theoretical contribution to the environmental governance literature, which is more attentive to the vertical and horizontal dimensions of governance within particular arenas of environmental policymaking and management (Francesch-Huidobro 2012).

Throughout this book, we devoted significant attention to the role of environmental movement and civil society players in shaping the oil-tourism interface. As others have noted, social movements studies focuses primarily on peak periods of mobilization and contention around specific issues, thereby creating a "distorted" understanding of social movement dynamics (McAdam et al. 2005). Rather than taking specific movements or controversies as our unit of analysis, we adopt a broader approach to the various ways in which movement players work in the multiple arenas of oil and tourism development. Movement players are engaged in a multiplicity of forms of action that involve conflict and collaboration with other players. Our results show that social movements exercise power through networks of contention and opposition, as is the focus of most social movement scholarship. However, we also see that movement players exercise power and generate social change through networks of collaboration with players across different sectors. These different forms of action also embody different tempos, from episodic and dramatic moments of anti-oil protest, to the everyday, less dramatic work of collaborating to build ecological literacy and social-ecological wellbeing at the community level. The players and arenas framework, which we used in our analysis, is well suited to this way of approaching social movement studies and offers a productive model for incorporating social movement dynamics into future political ecology research (Duyvendak and Jasper 2015; Jasper and Duyvendak 2015).

Though some government or oil sector players may at times view social movements as a nuisance, they play several important roles. They help shape public understandings and political orientations to oceans as natural resource pools, travel pathways, and sites of ecological problems and political contestation (Hannigan 2016; Widener 2018). They give voice to non-human nature and amplify tourism and fisheries interests where oil development creates risks to community and ecological well-being. As we see in the Norway case, movement players create friction against the expansion of fossil fuel development, but they also engage in collaboration around renewable energy transitions. Here, we see the value of movement players using diverse strategies of "conflictual cooperation" (Giugni and Passy 1998) with oil sector and government players depending on the specific issues at hand. Movement players also engage in projects with tourism players to help ensure that tourism can fulfil its potential as a site of environmental education. In our data, this orients around building ecological literacy for visitors to nature-oriented tourism attractions. This is consistent with other research that shows that outdoor adventure tourism and leisure helps promote environmental awareness and reflexivity about social-ecological relationships (Hanna et al. 2019;

Wheaton 2008). However, we ask whether there is potential to use environmentalism-tourism collaboration to nudge visitors even further in developing reflexivity about environmental problems and conflicts? This would help translate nature-oriented tourism into a more effective social force for provoking better informed ecological citizens, thereby fulfilling the potential for nature-based tourism to work as a space of environmental education and ecological literacy. Finally, social movement players can also help push the tourism sector to engage in reflexivity about the environmental costs of tourism, to adopt pro-environmental business practices around local ecological impacts, or to address the broader fossil fuel dependence of the sector. By focusing broadly on the multiple roles of social movement players in the oil-tourism interface across the North Atlantic, we make a valuable contribution to the social movements studies literature.

Lessons Learned: Climate Change and Social Futures

Climate change has been present throughout the book, though often at the fringes of the discussion, so now we bring it front and centre. Climate change is creating widespread ecological change across the North Atlantic region (Cunsolo and Ellis 2018; Rybråten et al. 2018; Shadian 2014). Environmental scientists speak of rapidly closing windows for substantial policy action and moves to decarbonization (Heede and Oreskes 2016). Environmental movements like the student strikes for climate or Fridays for the Future, Extinction Rebellion, and fossil fuel divestment are gaining media visibility and conveying the urgency of climate action through the language of climate crisis. Fossil fuel extraction, as well as the expansion of oil and gas infrastructure, are increasingly under attack by social movements. Protests over hydraulic fracturing ("fracking"), pipeline expansion, and Arctic oil exploration are a few examples in recent years (Cotton 2017; Kristoffersen and Dale 2014; Palmater 2015; Smith 2016; Stoddart et al. 2018). These campaigns are having an impact, with large institutions moving to divest from existing fossil fuel investments, or to withhold investment from new fossil fuel projects (Rowe et al. 2016; Stoddart et al. 2019). At the same time, renewable energy infrastructure, including solar power or onshore and offshore wind, are rapidly scaling up and changing the technological and economic landscape of our energy systems (Vasi 2011). In this rapidly changing terrain, tourism does not escape blame either. Environmentalist and media attention recently turned to the carbon cost of unnecessary air travel, leading to the notion of "flight shaming." The reliance of tourism on carbon-intensive air travel remains its most wicked climate change problem, as low-carbon technological solutions are not imminently on the horizon. Returning to the question that frames our analysis, "is co-existence possible between resource extraction and eco-tourisms," when we consider the context of climate change and the need for decarbonization, the answer becomes even more complex. Many host communities view the coexistence of the oil and tourism development as not only possible, but also *desirable* and socially beneficial. However, while co-existence is possible in the short-term, the long-term trajectories of climate change, policy action, and decarbonization makes it harder to see this version of the oil-tourism interface persisting over the long term.

Furthermore, the recent COVID-19 global pandemic has significantly disrupted both oil economies and tourism economies. The economic uncertainty and price volatility experienced by the oil sector since 2014 has been amplified by the crisis due to reduced travel and energy consumption. Oil briefly reached negative prices because of concerns about limited global storage capacity. At the same time, the hypermobility and rapid expansion of tourism that has been a hallmark of globalization since the 1990s has come to a jarring pause. COVID-19 has clearly revealed the "fragility" of depending on tourism as an economic driver for island and coastal societies (Sindico and Ellsmoor 2020). Communities that depend on oil or tourism are seeing significant negative economic impacts due to this global health pandemic. At the same time, social movement players have been making discursive connections between the public health and economic crises sparked by the COVID-19 outbreak, on one hand, and the climate crisis, on the other hand. This discourse frames the post-COVID recovery responses as an opportunity to pursue lower-carbon development pathways through the redesign of cities to prioritize active and public transportation, building retrofits, and green infrastructure. Conversely, movement players oppose emergency funding for oil and gas companies, airlines, and other carbon-intensive sectors. If movement players are successful in leveraging a post-COVID response as a pathway for decarbonization, this has implications for the future of both oil and tourism development.

Turning to lessons regarding the energy sector, we see that strong state involvement works in terms of maximizing social benefits for host communities and ensuring higher levels of environmental responsibility (also see Bryden 2015). Strong state involvement in the energy sector also builds capacity for envisioning and implementing new energy futures and grappling with the major social-ecological challenges of the twenty-first century. This includes taking a proactive and planned approach to renewable energy transitions, so that they are designed to be as socially beneficial as possible. This is particularly evident in Denmark, where public ownership in DONG Energy (now Ørsted) was leveraged to take an early leading role in wind power development. As oil price volatility has negatively impacted host communities, the Danish community of Esbjerg more successfully grappled with these challenges because of the scaling up of offshore wind power and positioning Esbjerg as a key hub for North Sea wind power installation. Furthermore, the Danish government is reconsidering the future of North Sea oil exploration and extraction in response to climate change concerns among the public (Reuters 2019). The ongoing expansion of wind energy is thus positioned as an environmental and economic solution to the need to decarbonize the Danish economy and energy system (Fleming 2019).

However, it is also important to note that renewable transitions are not free of tension and conflict. The development of wind, geothermal, or hydroelectric infrastructure produces new points of tension. For example, conflict over oil has subsided in Iceland, as exploration activity cooled in response to price declines and volatility. However, there have been tensions over renewable energy development, such as hydro and geothermal infrastructure. Opposition is grounded in concerns about the landscape impacts of these projects, which may degrade the wilderness values that help draw tourists and form a core part of Iceland's self-image for tourism promotion. There have also been conflicts between landowners in rural Denmark around the placement of onshore wind power, such that priorities for wind power development have shifted offshore. The Muskrat Falls hydroelectric project in Labrador has been framed as part of the provincial response to climate change. However, this project has provoked vocal opposition from the Labrador Land Protectors movement and members of Indigenous communities. There are concerns about slope stability issues and that the dam will create increased flows of methylmercury in the Churchill River, which will have significant downstream impacts on fish populations, with corresponding environmental health impacts for downstream communities that rely on traditional foods (Allen 2017).

These examples highlight that points of conflict also emerge during decarbonizing or shifting to renewable energy systems. These conflicts show that renewable energy transitions are not simply technological and economic issues but involve important social dynamics and considerations. As tensions around renewable energy highlights, there is no such thing as innocent energy. While the shift from fossil fuel-intensive energy systems to the increasing use of renewables is valuable and necessary, it is important to consider the social dimensions of renewable transitions as we think about building just and equitable energy futures in response to climate change (Jasanoff 2018).

Tourism—particularly in its ecotourism variants—is often seen by environmental groups and community members as a relatively sustainable alternative to the ecological risks inherent to fossil fuel production. Tourism is also seen as a potential contributor to meeting the UN Sustainable Development Goals (Gössling and Hall 2019; Hall 2019; Peeters et al. 2019). The UN World Tourism Organization (UNWTO) argues that "tourism has the potential to contribute, directly or indirectly to all of the goals," but particularly to SDG#8: decent work and economic growth; SDG#12: responsible consumption and production; and SDG#14: life below water (http://www2.unwto.org/content/tourism-2030-agenda). Responding to the carbon intensity of tourism travel, on one hand, and localized social and environmental impacts of overtourism, on the other hand, tourism scholars have begun to engage with ideas from degrowth movements. They argue that tourism development based on continuous economic growth and expanding tourism flows is fundamentally incompatible with ensuring the social and ecological wellbeing of host communities. From this perspective, the issue for host communities is not solely about growing tourism markets but developing alternative models of tourism development that prioritize more responsible forms of travel (Fletcher et al. 2019; Milano et al. 2019; Oklevik et al. 2019).

We have tried not to approach the oil-tourism interface with the assumptions that one of these development pathways is inherently good, while the other is inherently bad. As our results show, tourism is not an inherently benevolent or sustainable alternative to oil extraction. Tourism mobilities also depend on a variety of fossil fuel-based forms of travel, including airplane, car, bus, cruise ships, and boat tours. Plane travel is an especially difficult problem in treating tourism as a sustainable development pathway because of its carbon intensity and the lack of technological substitutes on the horizon (Cohen et al. 2011; Hanna and Adams 2019; Luzecka 2016). Solutions to the carbon intensity of tourism typically focus on changing individual tourists' practices to reduce the number and distance of flights, to focus on holidays that are closer to home, and to use land-based or public forms of transportation (Hall 2013; Iaquinto 2018; Smith et al. 2019). However, the cold-water tourism destinations of the North Atlantic are further from dense population centres and tourism markets, so depend on plane travel to connect global flows of visitors to local host communities. As such, sustainability solutions that rely on limiting long-haul tourism air travel may be unappealing in terms of being able to make tourism development economically sustainable for host communities. This highlights the social-ecological complexity of tourism as an alternative development pathway to resource extraction.

The lesson here is that we should not naively view tourism as a panacea for the problems inherent to resource extraction, or as something that stands outside climate change concerns. Rather, tourism travel is embedded within the causes, impacts, and needs to respond to climate change (Cohen et al. 2011; Hall 2010; Peeters et al. 2019; Scott et al. 2016). Moves to decarbonize tourism practices and decouple tourism from fossil fuel extraction are increasingly important. Our work identifies examples of this, such as Icelandic operators who are moving to the electrification of boat tours, or the electrification of the Hurtigruten coastal boat that is often used by tourists for transportation along coastal Norway. As societies grapple with climate change, these experiments will need to proliferate and become positive tipping points if tourism is to contribute to social futures characterized by social-ecological wellbeing.

Conclusion

Offshore oil and tourism are generally treated as separate development paths, though both are also seen to provide economic and social benefits for host communities. Our key concept of the oil-tourism interface and analysis of the contact points across sectors shows that they not only coexist but are intertwined in a variety of ways. The coexistence of these sectors is often taken for granted in host communities, until new projects for oil exploration and extraction impinge on established and valued tourism economies. This provokes conflict between different ways of valuing non-human nature and integrating it into social-ecological relationships and political economies. Periods of contention highlight the limitations within the oil-tourism interface that are often ignored. Approaching the oil-tourism interface through our relational political ecology perspective generates four major theoretical arguments.

First, nature-based tourism and oil extraction often share social-ecological space, but not cultural or political space. A comparative focus on evaluative repertoires and institutionalized orders of worth can help connect these two dimensions and shed light on the underlying conceptions of the common good that each development model speaks to (Boltanski and Thévenot 2006; Scott 2020; Ylä-Anttila and Luhtakallio 2016). When we compare evaluative repertoires across our cases, we find: (1) general talk of the environment frequently invokes the beauty of nature and is anchored in notions of inspirational worth; (2) oil development is primarily valued in terms of an industrial order of worth, and is secondarily valued in terms of a market order of worth; and (3) tourism development is evaluated in terms of an ecological order of worth, and secondarily by a market order of worth. While these associations are consistent across cases, we also find that the relative importance of each order of worth in each case is roughly homologous with the place of oil and tourism in the case's political economy. Also, the public discussion of these development models—in the form of news stories—is heavily siloed, representing "cultural holes" in how the oil-tourism interface is perceived, *and therefore governed*, in different societies (Pachucki and Breiger 2010). We argue that the concepts of evaluative repertoires and institutionalized orders of worth shed light on important differences within and across societies, thereby clarifying the cultural and political economic arenas that governance and social movement strategic action are embedded in.

Second, environmental governance literature attends to the vertical and horizontal dimensions of governance, which deal with the political scale (local, national, international) and range of actors involved in environmental policy arenas (Francesch-Huidobro 2012). However, we argue that "connective" dimensions of environmental governance need more attention. The connective dimension is about taking a more integrative and holistic look at relationships across different modes of development that share social-ecological space. This requires a shift in thinking about governance on the landscape/seascape level in terms of planning, policymaking, and environmental management. A major challenge to cultivating such connective forms of environmental governance is that coastal and ocean environments—which include landward, shoreline, nearshore, and offshore zones—are divided into separate political and managerial jurisdictions (Berkes 2015). Addressing this challenge will require a political openness to experimentation to institutionalize innovative and transformative changes in environmental governance, as well as a willingness to negotiate the power imbalances among different sectors and players (Gross 2010).

Third, we argue for greater attention to the multiple roles of environmental movements and civil society as players that provoke reflexivity and change within the oil-tourism interface. Environmental movement and civil society players also have the potential to help build social-ecological wellbeing as they raise questions about how we value our relationships with coastal and ocean environments (Hannigan 2016; Widener 2018). Much of the social movement studies literature isolates particular movements or issues of contention as case studies and objects of analysis. Instead, the players and arenas approach is better suited to grasping the spectrum of action and projects in which movement players engage (Duyvendak and Jasper 2015; Jasper and Duyvendak 2015). In the oil-tourism interface, this involves projects of collaboration and conflict with other players, including government, the energy sector, the tourism sector, and others. Environmental movement strategic interaction also involves heightened moments of contention, as well as more mundane projects for sustainability that would otherwise pass unnoticed by models of social movement scholarship that focus on peak periods of conflict.

Finally, climate change is substantially transforming the social-ecological context of the North Atlantic. This has significant implications for the oil-tourism interface, as well as for fisheries and community infrastructure. For many host communities, the co-existence of resource extraction and ecotourism is acceptable and desirable, other than in specific episodes of conflict. However, in the context of climate change, there needs to be an increased decoupling of ecotourisms from fossil fuel resource extraction and moves to decarbonization in both sectors. This energy transition and transformation of the oil-tourism interface will come with new points of tension and conflict, as communities face the uneven distributions of costs and benefits of decarbonization and renewable energy development. Moving forward, this requires an analytical shift from the theoretical framework of the oil-tourism interface to a broader *energy*-tourism interface that continues to navigate the complexities of the multiple contact points between energy systems and tourism travel as coastal societies grapple with creating social futures that ensure social-ecological wellbeing.

References

Allen, V. (2017). Muskrat Falls. In L. Moore & A. Marland (Eds.), *The democracy cookbook: Recipes to renew governance in Newfoundland and Labrador* (pp. 318–320). St. John's, NL: ISER Books.

Bennett, J. (2010). *Vibrant matter: A political ecology of things*. Durham, NC: Duke University Press.
Berkes, F. (2015). *Coasts for people: Interdisciplinary approaches to coastal and marine resource management*. New York: Routledge.
Boltanski, L., & Thévenot, L. (2006). *On justification: Economies of worth* (C. Porter, Trans.). Princeton, NJ: Princeton University Press.
Bravo, M. T. (2009). Voices from the sea ice: The reception of climate impact narratives. *Journal of Historical Geography, 35*, 256–278.
Bryden, J. (2015). The development of industry and North Sea oil in Scotland and Norway. In J. Bryden, O. Brox, & L. Riddoch (Eds.), *Northern neighbours: Scotland and Norway since 1800* (pp. 123–153). Edinburgh, UK: Edinburgh University Press.
Callison, C. (2014). *How climate change comes to matter: The communal life of facts*. Durham, NC: Duke University Press.
Campbell, C. E. (2019). The wealth of wilderness. In C. M. Coates & G. Wynn (Eds.), *The nature of Canada* (pp. 167–184). Vancouver, BC: On Point Press.
Cohen, S. A., Higham, J. E. S., & Cavaliere, C. T. (2011). Binge flying: behavioural addiction and climate change. *Annals of Tourism Research, 38*(3), 1070–1089.
Cotton, M. (2017). Fair fracking? Ethics and environmental justice in United Kingdom shale gas policy and planning. *Local Environment, 22*(2), 185–202.
Crossley, N. (2010). *Towards relational sociology*. London: Routledge.
Cunsolo, A., & Ellis, N. R. (2018). Ecological grief as a mental health response to climate change-related loss. *Nature Climate Change, 8*(4), 275.
Dale, B. (2016). Governing resources, governing mentalities: Petroleum and the Norwegian integrated ecosystem-based management plan for the Barents and Lofoten seas in 2011. *The Extractive Industries and Society, 3*(1), 9–16.
Dale, B., & Kristoffersen, B. (2018). Post-petroleum security in a changing Arctic: Narratives and trajectories towards viable futures. *Arctic Review on Law and Politics, 9*, 244–261.
Davidson, D. J., & Stedman, R. C. (2018). Calling forth the change-makers: Reflexivity theory and climate change attitudes and behaviors. *Acta Sociologica, 61*(1), 79–94.
Duyvendak, J. W., & Jasper, J. M. (2015). *Breaking down the state: Protestors engaged*. Amerstdam: Amsterdam University Press.
Escobar, A. (1999). After nature: Steps to an antiessentialist political ecology. *Current Anthropology, 40*(1), 1–30.

Fleming, S. (2019). *Wind farms now provide 14% of EU power: These countries are leading the way*. https://www.weforum.org/agenda/2019/03/wind-farms-now-provide-14-of-eu-power-these-countries-are-leading-the-way/. Accessed 19 November 2019.

Fletcher, R., Mas, I. M., Blanco-Romero, A., & Blázquez-Salom, M. (2019). Tourism and degrowth: An emerging agenda for research and praxis. *Journal of Sustainable Tourism, 27*(12), 1745–1763.

Francesch-Huidobro, M. (2012). Institutional deficit and lack of legitimacy: The challenges of climate change governance in Hong Kong. *Environmental Politics, 21*(5), 791–810.

Giugni, M. G., & Passy, F. (1998). Contentious politics in complex societies: New social movements between conflict and cooperation. In M. G. Giugni, D. McAdam, & C. Tilly (Eds.), *From contention to democracy* (pp. 81–108). Lanham, MD: Rowman & Littlefield.

Gössling, S., & Michael Hall, C. (2019). Sharing versus collaborative economy: how to align ICT developments and the SDGs in tourism? *Journal of Sustainable Tourism, 27*(1), 74–96.

Gould, K. A. (2017). Ecotourism under pressure: The political economy of oil extraction and cruise ship tourism threats to sustainable development in Belize. *Environmental Sociology, 3*(3), 237–247.

Gross, M. (2010). *Ignorance and surprise: Science, society, and ecological design*. Cambridge, MA: The MIT Press.

Hall, C. M. (2010). Tourism and environmental change in polar regions: Impacts, climate change and biological invasion. In C. M. Hall & J. Saarinen (Eds.), *Tourism and change in polar regions: Climate, environments and experiences* (pp. 42–70). London: Routledge.

Hall, C. M. (2013). Framing behavioural approaches to understanding and governing sustainable tourism consumption: Beyond neoliberalism, "nudging" and "green growth"? *Journal of Sustainable Tourism, 21*(7), 1091–1109.

Hall, C. M. (2019). Constructing sustainable tourism development: The 2030 agenda and the managerial ecology of sustainable tourism. *Journal of Sustainable Tourism, 27*(7), 1044–1060.

Hall, C. M., Müller, D. K., & Saarinen, J. (2009). *Nordic tourism: Issues and cases*. Bristol, UK: Channel View Publications.

Hall, P. A., & Soskice, D. (2001). An introduction to varieties of capitalism. In P. A. Hall & D. Soskice (Eds.), *Varieties of capitalism: The institutional foundations of comparative advantage* (pp. 1–70). Oxford, UK: Oxford University Press.

Hanna, P., & Adams, M. (2019). Positive self-representations, sustainability and socially organised denial in UK tourists: discursive barriers to a sustainable transport future. *Journal of Sustainable Tourism, 27*(2), 189–206.

Hanna, P., Wijesinghe, S., Paliatsos, I., Walker, C., Adams, M., & Kimbu, A. (2019). Active engagement with nature: outdoor adventure tourism, sustainability and wellbeing. *Journal of Sustainable Tourism, 27*(2), 189–206.

Hannigan, J. (2016). *The geopolitics of deep oceans*. Cambridge, UK: Polity Press.

Heede, R., & Oreskes, N. (2016). Potential emissions of CO2 and methane from proved reserves of fossil fuels: An alternative analysis. *Global Environmental Change, 36,* 12–20.

Iaquinto, B. L. (2018). Backpacker mobilities: Inadvertent sustainability amidst the fluctuating pace of travel. *Mobilities, 13*(4), 569–583.

James, L., Olsen, L. S., & Karlsdóttir, A. (2020). Sustainability and cruise tourism in the arctic: Stakeholder perspectives from Ísafjörður, Iceland and Qaqortoq, Greenland. *Journal of Sustainable Tourism, 28*(9), 1425–1441.

Jasanoff, S. (2018). Just transitions: A humble approach to global energy futures. *Energy Research & Social Science, 35,* 11–14.

Jasper, J. M., & Duyvendak, J. W. (Eds.). (2015). *Players and arenas: The interactive dynamics of protest*. Amsterdam: Amsterdam University Press.

Jenkins, K., McCauley, D., Heffron, R., Stephan, H., & Rehner, R. (2016). Energy justice: A conceptual review. *Energy Research & Social Science, 11,* 174–182.

Jennings, A. (2015). The old rock's rockin': A cultural anatomy of Shetland. In K. Vodden, R. Gibson, & G. Baldacchino (Eds.), *Place peripheral: Place-based development in rural, island, and remote regions* (pp. 156–179). St. John's, NL: ISER Books.

Kristoffersen, B., & Dale, B. (2014). Post petroleum security in Lofoten: How identity matters. *Arctic Review on Law and Politics, 5*(2), 201–226.

Latour, B. (2004). Politics of nature: How to bring the sciences into democracy (C. Porter, Trans.). Cambridge, MA: Harvard University Press.

Le Billon, P., & Kristoffersen, B. (2019). *Just cuts for fossil fuels? Supply-side carbon constraints and energy transition.* Environment and Planning A: Economy and Space. https://doi.org/10.1177/0308518X18816702.

Lijphart, A. (2012). *Patterns of democracy: Government forms and performance in thirty-six countries* (2nd ed.). New Haven, CT: Yale University Press.

Longo, S. B., & Clark, B. (2016). An ocean of troubles: Advancing marine sociology. *Social Problems, 63,* 463–479.

Lund, K. A., Loftsdóttir, K., & Leonard, M. (2017). More than a stopover: Analysing the postcolonial image of Iceland as a gateway destination. *Tourist Studies, 17*(2), 144–163.
Luzecka, P. (2016). "Take a gap year!" A social practice perspective on air travel and potential transitions towards sustainable tourism mobility. *Journal of Sustainable Tourism, 24*(3), 446–462.
McAdam, D., Sampson, R., Weffer, S., & MacIndoe, H. (2005). 'There will be fighting in the streets': The distorting lens of social movement theory. *Mobilization: An International Quarterly, 10*(1), 1–18.
McCauley, D. (2018). Reframing decommissioning as energy infrastructural investment: a comparative analysis of motivational frames in Scotland and Germany. *Energy Research & Social Science, 41*, 32–38.
Milano, C., Novelli, M., & Cheer, J. M. (2019). Overtourism and degrowth: A social movements perspective. *Journal of Sustainable Tourism, 27*(12), 1857–1875.
Motion, J. (2019). Undoing art and oil: An environmental tale of sponsorship, cultural justice and climate change controversy. *Environmental Politics, 28*(4), 727–746.
Oklevik, O., Gössling, S., Hall, C. M., Jacobsen, J. K. S., Grøtte, I. P., & McCabe, S. (2019). Overtourism, optimisation, and destination performance indicators: A case study of activities in fjord Norway. *Journal of Sustainable Tourism, 27*(12), 1804–1824.
Pachucki, M. A., & Breiger, R. L. (2010). Cultural holes: Beyond relationality in social networks and culture. *Annual Review of Sociology, 36*, 205–224.
Palmater, P. (2015). *Indigenous nationhood*. Halifax: Fernwood.
Peeters, P., Higham, J., Cohen, S., Eijgelaar, E., & Gössling, S. (2019). Desirable tourism transport futures. *Journal of Sustainable Tourism, 27*(2), 173–188.
Ramos, H., Stoddart, M. C., & Chafe, D. (2016). Assessing the tangible and intangible benefits of tourism: Perceptions of economic, social, and cultural impacts in Labrador's Battle Harbour Historic District. *Island Studies Journal, 11*(1), 209–226.
Reuters. (2019). Denmark weighs ending North Sea oil and gas hunt. Retrieved from https://www.reuters.com/article/us-denmark-oil/denmark-weighs-ending-north-sea-oil-and-gas-hunt-idUSKBN1WQ1J8. Accessed 19 November 2019.
Rowe, J., Dempsey, J., & Gibbs, P. (2016). The power of fossil fuel divestment (and its secret). In W. K. Carroll & K. Sarker (Eds.), *A world to*

win: Contemporary social movements and counter-hegemony (pp. 233–249). Winnipeg, MB: ARP Books.

Rybråten, S., Bjørkan, M., Hovelsrud, G. K., & Kaltenborn, B. P. (2018). Sustainable coasts? Perceptions of change and livelihood vulnerability in Nordland, Norway. *Local Environment, 23*(12), 1156–1171.

Scott, D., Hall, C. M., & Gössling, S. (2016). A report on the Paris climate change agreement and its implications for tourism: Why we will always have Paris. *Journal of Sustainable Tourism, 24*(7), 933–948.

Scott, N. A. (2020). *Assembling moral mobilities: Cycling, cities, and the common good*. Lincoln, Nebraska: University of Nebraska Press.

Shaban, H. (2019). *He paid Wow Air $201 to get to Iceland: Then he got stuck*. https://www.washingtonpost.com/business/2019/03/28/wow-air-abruptly-shuts-down-stranding-passengers/. Accessed 22 January 2020.

Shadian, J. M. (2014). *The politics of Arctic sovereignty: Oil, ice, and Inuit governance*. London: Routledge.

Sindico, F., & Ellsmoor, J. (2020). *COVID-19: A global island response*. Strathclyde, UK: University of Strathclyde, Island Innovation.

Smith, A., Robbins, D., & Dickinson, J. E. (2019). Defining sustainable transport in rural tourism: Experiences from the New Forest. *Journal of Sustainable Tourism, 27*(2), 258–275.

Smith, J. R. (2016). *When petro-capitalism comes knocking: Community interpretations and responses to the Gros Morne fracking controversy* (MA thesis). St. John's, NL: Memorial University of Newfoundland.

Sovacool, B. K., Burke, M., Baker, L., Kotikalapudi, C. K., & Wlokas, H. (2017). New frontiers and conceptual frameworks for energy justice. *Energy Policy, 105*, 677–691.

Statistics Iceland. (2019). *Tourism short-term indicators in July*. https://www.statice.is/publications/news-archive/tourism/tourism-short-term-indicators-in-july/. Accessed 22 January 2020.

Stoddart, M. C. J. (2012). *Making meaning out of mountains: The political ecology of skiing*. Vancouver, BC: University of British Columbia Press.

Stoddart, M. C. J., & Graham, P. (2018). Offshore oil, environmental movements, and the oil-tourism interface: the Old Harry conflict on Canada's east coast. *Sociological Inquiry, 88*(2), 274–296.

Stoddart, M. C., Chewinski, M., Burt, B. Q., & Stewart, M. (2019). Political consumerism in the oil and mining extractive industries: Possibilities for sustainability and social justice. In M. Bostrom, M. Micheletti, & P. Oosterveer (Eds.), *The Oxford handbook of political consumerism* (pp. 367–388). New York: Oxford University Press.

Stoddart, M. C., Smith, J. R., & Graham, P. (2018). Oil opposition. In D. J. Davidson & M. Gross (Eds.), *The Oxford handbook of energy and society* (pp. 447–460). New York: Oxford University Press.

Urry, J. (2014). *Offshoring*. Cambridge, UK: Polity Press.

Vasi, I. B. (2011). *Winds of change: The environmental movement and the global development of the wind energy industry*. New York: Oxford University Press.

VisitNorway. (2019). *The joy of different seasons*. https://www.visitnorway.com/plan-your-trip/seasons-climate/. Accessed 28 January 2020.

Wheaton, B. (2008). From the pavement to the beach: Politics and identity in 'Surfers Against Sewage'. In M. Atkinson & K. Young (Eds.), *Tribal play: Subcultural journeys through sport* (pp. 113–134). Bingley, UK: JAI Press.

Widener, P. (2011). *Oil injustice: Resisting and conceding a pipeline in Ecuador*. Lanham, MD: Rowman & Littlefield Publishers.

Widener, P. (2018). Coastal people dispute offshore oil exploration: Toward a study of embedded seascapes, submersible knowledge, sacrifice, and marine justice. *Environmental Sociology, 4*(4), 405–418.

Ylä-Anttila, T., & Luhtakallio, E. (2016). Justifications analysis: understanding moral evaluations in public debates. *Sociological Research Online, 21*(4), 1–14.

Young, M., Markham, F., Reis, A. C., & Higham, J. E. S. (2015). Flights of fantasy: A reformulation of the flyers' dilemma. *Annals of Tourism Research, 54,* 1–15.

7

Epilogue on Methodology

Methodological Standpoint: Omnivorous, Pragmatic, Mobile

Issues related to social-ecological systems and environmental governance are complex, in terms of the increasingly diverse range of players and multiple political scales from the local to the global, in terms of the intersections of social and environmental science, and in terms of governance and environmental justice. Gaining a better understanding of this complexity demands a diverse range of methodological tools and approaches (Small and Adler 2019). As noted in Chapter 1, this leads us to embrace a standpoint of "methodological omnivorousness," which offers an intellectual co-working space across qualitative and quantitative methods, with their different emphases on narrative and numerical data. As such, our methodological standpoint for this project is characterized by three principles: methodological omnivorousness, pragmatism, and mobilities. Below, we clarify what we mean by each of these principles.

Being a methodological omnivore does not mean that every researcher needs to become an expert in qualitative and quantitative methods. Instead, we suggest this is about an openness to the value of methods beyond one's own expertise, as well as recognizing the importance of

M. C. J. Stoddart et al., *Industrial Development and Eco-Tourisms*,
https://doi.org/10.1007/978-3-030-55944-1_7

working in collaboration with colleagues who have diverse methodological interests and expertise. Theoretically and methodologically diverse teams are more capable than any individual collaborator or method (Page 2019; Graham et al. 2020). In the case of our own collaboration, Stoddart approaches this project primarily as a qualitative researcher with experience in interviewing, fieldwork, and textual analysis, and with moderate experience with social network analysis and survey research. Mattoni also comes to the project primarily as a qualitative researcher with an emphasis on fieldwork and interviewing, media analysis and visual sociology, and comparative analysis. McLevey, by contrast, is a mixed-methods researcher whose primary expertise is in computational social science, including social network analysis and applied natural language processing, and secondarily in survey methodology and qualitative interviewing. We see our collaboration as extending the range of possibilities to generate more complex analyses and research accounts. Methodological omnivorousness is more about playing well with others who have different skill sets and is emphatically not about trying to be a jack of all trades, but master of none (Stokes 2020).

The notion of methodological omnivorousness is complementary to work by Laurel Richardson (2000) and John Law (2004) that attempts to identify the value of mixing methodological approaches beyond simply "triangulating" findings. These alternative approaches emphasize how expanding our methodological toolkit can lead to more poly-vocal and complex research accounts. Writing from a primarily qualitative perspective, Richardson (2000) offers the metaphor of crystallization, which uses the image of how facets of a crystal refract light in different ways. Similarly, Law (2004) draws on the sociology of science literature to encourage the use of multiple methods to better grasp the "messiness" of social life. We also see methodological omnivorousness as complimentary to the range of relational sociologies, including the relational political ecology framework used throughout our analysis.

The methodological dividing line between qualitative and quantitative methodological communities is often firm and difficult to permeate, in part because they are embedded in different epistemic and evidential cultures (Lamont 2009; Collins 2004; McLevey 2019). As Alford (1998) argues, while all projects have at least one foreground method (quantitative, qualitative, or historical), the others are at least implicit. While interest in mixed-methods research is growing, many researchers assert that there are incommensurable differences in the core operating assumptions of qualitative and quantitative research paradigms (Morgan 2007). We prefer David Morgan's (2007) perspective of methodological pragmatism as a solution, which he describes as follows:

> A far more interesting option is to explore the potential for working back and forth between the kinds of knowledge we have already produced under the separate banners of Qualitative and Quantitative Research. What if Quantitative Researchers paid more attention to the incredible range of hypotheses that Qualitative Researchers have "generated" for them? And what if Qualitative Researchers spent more time exploring the range of phenomena that Quantitative Researchers have sought to define and test? Rather than each camp dismissing the others' work as based on wholly incompatible assumptions, our goal would be to search for useful points of connection. These are the kinds of opportunities that a pragmatic approach to social science research has to offer. (Morgan 2007, p. 71)

There are various approaches to mixed-methods research design (Small 2011). One approach is to conceptualize mixed-methods work in terms of priority and sequence (Alford 1998; Cresswell 2003). Priority refers to which methods are given prominence in a project. Mixed-method research design can be qualitative dominant, quantitative dominant, or symmetrical. Sequence refers to the order of approaches within a project, in terms of whether the qualitative or quantitative phase comes first, or whether both phases proceed in parallel. Furthermore, mixing qualitative and quantitative modes of research can happen at several points in a project, including data collection, data analysis, and in writing and publication.

At the level of data collection, we can characterize the research design for this project as a quant → QUAL/quant approach. Structured online interviews occurred first, followed by qualitative fieldwork and interviews. Other data collection strategies proceeded in parallel with this, including quantitative mass media corpus analysis, qualitative analysis of online content (organizational websites and social media content), and other document analysis. Despite starting with quantitative methods, qualitative data are dominant in the project overall, with the quantitative analysis taking a supportive role. However, we also mixed qualitative and quantitative modes at the level of analysis. For example, the discourse network figures used in the book are constructed by generating coding matrices of co-occurrences of themes and actors in our qualitative data, and from semi-supervised computational models that explicitly incorporate findings from the qualitative components of the project into a more quantitative framework. In different ways, these processes involve quantifying our qualitative data. This affords a variety of benefits, such as making it possible to identify the relational dimensions of the oil-tourism interface by visualizing network structure.

The fieldwork for this project also draws on mobile methodologies. Following the theoretical and conceptual innovations of the "mobilities turn," researchers have developed a range of methods to better follow, document, and analyse mobilities (Blok 2010; Scott 2020; Spinney 2011; Thorpe 2014). The edited collection oriented around *Mobile Methods* is a key work for articulating what is unique and important about these methods. In their introduction to the volume, Büscher et al. (2011b) note that mobilities research focuses on five types of mobility that are increasingly prevalent. These include: "*corporeal travel* of people for work, leisure, family file, pleasure, migration and escape"; "*physical movement* of objects to producers, consumers and retailers"; "*imaginative travel*" involving the consumption of mediated images of "places and peoples appearing on and moving across multiple print and visual media"; "*virtual travel* … that enables presence and action at a distance"; and "*communicative travel* through person-to-person contact" by communication technologies including messages, texts, or mobile calls (Büscher et al. 2011b, p. 5). From this, Büscher et al. note that mobilities research methods are based on the premise that: "Researchers will benefit if they capture, track, simulate, mimic and shadow the many and interdependent forms of intermittent movement of people, images, information and objects…" (2011b, p. 7). In the conclusion to this volume, Büscher et al. further assert the promise of mobile methods, claiming, "Mobile methods are good at knitting together the different material, technical, social, temporal and spatial scales at which everyday actions take effect" (Büscher et al. 2011a, p. 121).

There are several benefits of working from a methodological standpoint that foregrounds movement and mobility. If we accept that contemporary social worlds are characterized by the intensification of local-global connectivity and movement, then it makes sense that researchers should follow these flows of movement and engage in research on the move with participants, or across multiple research sites. Our project uses multiple forms of data collection and analysis that combine "field-based" and "desk-based" approaches (Merriman 2014). However, fieldwork is at the core of the project, carried out through mobile fieldwork at sites around the North Atlantic, including Lofoten Islands, Stavanger, Oslo, and Bergen (Norway); Aberdeen, Glasgow, Edinburgh, and Aviemore (Scotland); Reykjavik and Akureyri (Iceland); Esbjerg, Ribe, and Copenhagen (Denmark). The project also extends several years of research carried out in Newfoundland and Labrador, which includes fieldwork in St. John's and western Newfoundland (including Gros Morne National Park), as well as the Burin Peninsula and Labrador Straits regions. The mobile approach to multi-sited field research allowed us to draw connections across sites related to offshore oil and nature tourism across the North Atlantic.

Data Collection

The focus of this project on the oil sector and nature-based tourism is driven by theoretical concerns and practical considerations. From a theoretical perspective, both the oil and tourism sectors have been key to political and public discourse about economic diversification and development at different periods across our case study regions. For example, both sectors have been positioned as pathways for social and economic recovery for coastal societies that faced the decline and restructuring of their fishing industries. From a practical perspective, the project scope and data collection were constrained by available resources. As such, we focused data collection on these two forms of development across the five case study regions, with field notes, interviews, and textual analysis focusing on the oil sector, tourism sector, government, and environmental movements. Of course, other key industries also share social-ecological space with offshore oil and tourism, most notably fisheries and shipping. However, due to research capacity and resources, we have chosen to narrow our focus onto oil and tourism as two sectors and development pathways that provide theoretically rich ground for comparison.

Similarly, we struggled with whether and how to include Greenland within the project scope. Greenland is an important region to include in comparative research on social-ecological wellbeing in North Atlantic. Greenland has an emerging tourism sector that is shaped by visitors' desires to see Arctic landscapes before they are reconfigured by climate change (fieldnotes, October 8–9, 2016; also see James et al. 2020; Ren et al. 2016). At the same time, interest in offshore oil exploration based in Greenland has ebbed and flowed in response to global oil prices. For many Greenlanders, including Inuit community leaders, these forms of development are viewed positively insofar as they can be leveraged to gain greater independence from the Kingdom of Denmark (Nuttall 2012; Shadian 2014). Ideally, we would have included Greenland as an additional case study region. Unfortunately, project resources did not allow for the fieldwork necessary to ensure comparable data with our other case study regions. We also did not want to subsume Greenland within the Danish case. Our imperfect solution has been to discuss Greenland as relevant in the broader contextual chapters of the book (Chapters 1 and 2), but to limit discussion elsewhere. Extending our analysis with a deeper examination of the Greenlandic case is an important area for future research.

The project combines several modes of data collection. First, structured interviews were carried out in 2016 using an online instrument ($n = 46$). We constructed a sampling frame of 341 key organizations from the oil sector, tourism sector, relevant government agencies, and environmental organizations. The overall response rate was only 13%, though this varied significantly across cases (better for Newfoundland and Labrador than the other cases), as well as across sectors (better for tourism sector and environmental organizations, but worst for the oil sector). Questions focused on beliefs about the economic and social significance of the oil and tourism sectors, as well as the perceived environmental sustainability of the sectors. Questions also asked about relationships of collaboration and conflict within and across the sectors. Given the overall low response rate and clear evidence of non-response bias, we do not rely on these results in our data analysis.

Fieldwork was also carried out across our case study regions, resulting in 119 days of observational data. This included "event ethnography" of several industry conferences and other events. Whereas traditional forms of ethnography focus on long-term immersion in a place, event ethnography focuses on short-term, temporarily bound events that create temporary social arenas (Campbell et al. 2014). A major advantage of event ethnography is that it allows researchers to access spaces of discussion among members of economic and political elite organizations that are difficult to recruit for surveys or interviews. Event ethnography provides insight and reveals themes that are not apparent in official policy documents, websites, or other publicly available data.

Fieldwork was conducted in Norway (oriented around Lofoten Islands, Stavanger, and Oslo) from August 17 to September 16, 2016, which included event ethnography at Offshore North Seas, the major annual Norwegian oil sector conference held in Stavanger. A follow-up fieldwork trip in Bergen took place from September 10–15, 2017. Fieldwork was conducted in Iceland (oriented around Reykjavik and Akureyri) from October 4–29, 2016, with a follow-up trip to Reykjavik from October 17–23, 2018. This included event ethnography at the 2016 and 2018 Arctic Circle conferences, which are attended by a mix of policymakers and practitioners, community leaders, researchers, and NGO representatives. This event covers a wide range of Arctic-related issues, including tourism, natural resource development, and climate change. Fieldwork was conducted in Scotland (oriented around Aberdeen, Glasgow, Aviemore, and Edinburgh) from March 6 to April 1, 2017. This included event ethnography at the Oil & Gas UK 2017 Outlook Report launch event, as well as the 2017 Scottish Hospitality and Tourism Week conference in Glasgow and Wild Scotland conference in Aviemore. Fieldwork was conducted in Denmark (oriented around Esbjerg, Ribe, and Copenhagen) from April 17 to May 12, 2017. Principle investigator Stoddart is based at Memorial University in St. John's, so fieldwork in Newfoundland and Labrador unfolded over a longer period from June 21, 2016 to July 3, 2018. Fieldwork oriented primarily around St. John's and Gros Morne National Park. This included event ethnography at the 2016 and 2017 NOIA (Newfoundland and Labrador Offshore Oil and Gas Industries Association) conferences, as well as the 2018 Hospitality Newfoundland and Labrador conference.

As part of the fieldwork, we conducted 62 semi-structured interviews with participants from the energy sector, tourism operators and stakeholders, environmental organizations, and various relevant local, regional, and national government offices. Recruitment for these qualitative interviews was based on the same sampling frame used for the online structured interviews. Table 7.1 is a sampling matrix that breaks down the sample by region and sectoral affiliation of participants. As the table indicates, oil sector participants were under-represented compared with other groups, highlighting the importance of using a mixed-method approach with multiple data sources.

The interviews were semi-structured, with an interview schedule with set questions for all participants. However, we diverged from the schedule as necessary to follow-up on interesting points, or re-ordered questions to respond to the flow of conversation. The interview was broken into four sections. The first set of questions delved into the positive and negative impacts of nature-based tourism, including issues of environmental sustainability. The second section posed a parallel set of questions about the offshore oil industry. This section included additional questions about the relationships across the tourism industry and oil industry, as well as about the relative importance of tourism and offshore oil

Table 7.1 Interview sampling matrix

	Newfoundland and Labrador	Norway	Iceland	Denmark	Scotland	Total
Tourism industry	5	2	4	5	3	19
Energy sector	1	0	2	2	2	7
Government	3	6	3	6	6	24
Environmental organizations	7	3	1	1	0	12
Total	16	11	10	14	11	62

as economic drivers. The third section focused on social network relationships of collaboration and conflict related to tourism and offshore oil within and across sectors. The fourth section asked questions about participants' engagement in media-work related to tourism or oil. These last questions were treated as optional depending on whether they were relevant based on participants' roles in their organizations, as well as their available time. Interviews were conducted in person wherever possible, in locations chosen by participants, and by telephone when necessary. Interviews ranged from 30 to 90 minutes in duration, with most lasting 45–60 minutes.

We also carried out web-based textual analysis, which has been referred to as "internet ethnography" or "netnography" (Hine 2005; Small and Harris 2014). This involves treating the internet as a site for field observation. We took a hybrid approach to sampling web content from tourism, energy sector, government, and environmental organization websites. We identified key organizations through online keyword searching and supplemented this with organizations that were identified through interviews and field observation. A field note protocol was used to guide notetaking on each of the websites, which was carried out by graduate student research assistants. We analysed 99 websites in total, as described in the sampling matrix in Table 7.2. The netnography protocol prompted for notes on the following: dominant themes and images used throughout the websites to represent coastal environments and communities; if and how the language of environmental sustainability was used; representations of positive and negative impacts of tourism for host communities; representations of positive and negative

Table 7.2 Website sampling matrix

	Newfoundland and Labrador	Norway	Iceland	Denmark	Scotland	Total
Tourism industry	5	4	8	8	4	29
Energy sector	8	14	2	3	8	35
Government	6	3	2	1	5	17
Environmental Organizations	1	4	5	5	3	18
Total	20	25	17	17	20	99

impacts of offshore oil for host communities; climate change impacts and responses; and relationships of collaboration and/or conflict across sectors. We also attended to how the websites represented people in terms of social markers of class, gender, race and ethnicity, or sexual orientation. We also gathered information on links to other websites, including social media platforms.

We gathered additional data on social movements' organizations through four social media platforms: Facebook, Instagram, Twitter, and YouTube. We analysed 85 social media profiles in total, gathering posts from these platforms as detailed in Table 7.3. We started from the social movement organizations that the previous round of qualitative analysis identified as involved in the oil and tourist sectors in each case study region. For each of them, we looked for their official profiles—or pages, groups, and channels according to the social media platform at stake—and gathered their posts from the past five years. Data collection occurred manually and was focused on content that directly referred to either the oil sector or the tourism sector. The resulting analysis included, for each social movement organization, three sections: key discourses that emerge in posts where the social movement organization intervened with the oil and/or tourism industry; main forms of collective action, both mediated and non-mediated, that the social movement organization used through its social media channels; main relationships of collaboration and conflict between the social movement organization and other actors involved in the oil and/or tourism sector.

We also engaged in a document analysis of 72 policy documents and other relevant documents dealing with tourism and offshore oil that were produced by government, industry associations, or other key

Table 7.3 Social movement social media sampling matrix

	Newfoundland and Canada	Scotland	Norway	Denmark	Iceland	Total
Twitter	5	4	6	5	7	27
Facebook	7	4	5	4	6	26
YouTube	3	3	3	4	3	16
Instagram	2	2	5	5	2	16
Total	17	13	19	18	18	85

organizations across our case study regions. Several of these documents were recommended by participants during interviews or were referenced during event ethnographic fieldwork.

Finally, we collected data on a corpus of news stories, with the goal of identifying stories that spoke to either oil development, tourism development, or both. We searched the Factiva major newspaper database for English-language news articles that contained both the words "tour*" and "oil" (or "petroleum") as well as the name(s) of at least one of our case study regions. These stories were published by major mainstream media sources in either Europe, the United States, Canada, or the U.K. over 6 years between 1 January 2011 and 31 December 2016.

Our relatively open search strategy returned many irrelevant articles, so we manually screened the results and removed irrelevant articles from the corpus. Next, we coded each remaining article as being about one or more of the case study regions. Our final corpus consisted of 1,418 articles, 848 of which were about Scotland, 276 about Norway, 135 about Newfoundland and Labrador, 92 about Iceland, and 67 about Denmark. To facilitate case comparisons, we constructed separate corpuses for each case.

Data Analysis

A project management file was created in NVIVO software for qualitative analysis, where most of the coding and analysis was done. All original field notes, interview transcripts, netnography notes, and documents were stored in NVIVO and assigned attributes based on case study region and sector. Qualitative coding was semi-structured, beginning with a theoretically informed coding scheme, but allowing this to evolve in response to emerging themes from the data. Top-level thematic categories, which were subdivided into more precise second-level categories, included: culture and collective identity; ecological dimensions; media communication; mobilities; social power; political dimensions; social futures; social network dynamics; and socio-economic dimensions. We also coded all mentions of organizational actors in our data, organized as follows: energy sector; environmental movement opponents;

environmental organizations and civil society; government and political parties; international agencies; media; tourism; and universities and research centres.

Quantitative data from our structured interviews was analysed separately. Due to concerns with low response rates and non-response bias, we treated this analysis as *exploratory*. Our exploratory analysis of this data suggested interesting patterns that we examined more deeply and systematically in the interviews, fieldwork, and other data. This included the finding that there is a generally positive view of tourism, but that views about the oil sector are more diverse and contested. This also included the finding that where conflict emerges in the oil-tourism interface, it typically involves alignments of oil sector and government players against tourism and environmental organization players. The summary report from this phase of the project was imported to the master NVIVO file so that it could be linked to the other data sources.

Similarly, the media corpus analysis was carried out separately. This work proceeded in three stages: (1) preparation and cleaning, (2) analysis, and (3) integrating the findings with our other data analyses.

First, we performed all recommended pre-processing steps using the Python package spaCy (Honnibal and Montani 2017). For each article, we removed English-language stopwords ("and," "of," "the," etc.) and converted all text to lower case. We used part-of-speech tagging to extract a list of nouns used in each article, including multi-word phrases (e.g. "climate_change"). Focusing on nouns strips information from individual articles but improves the ability of our analysis to identify coherent topics. This trade-off is well-understood in the field of natural language processing and computational text analysis (Bail 2016; Rule et al. 2015). We normalized words in the corpus by lemmatization (Grimmer and Stewart 2013; Ignatow and Mihalcea 2017; McLevey 2021), which reduces words to their base dictionary form and combines words with superficial differences. Lemmatization is superior to alternatives because it returns valid words that are easier to interpret and preserves as much information as possible.

We used topic models for data analysis, which in recent years have been widely employed partly because of their affinities with relational theories of meaning and language and their usefulness for looking at the big picture of cultural environments and political deliberation (Blei 2012; DiMaggio et al. 2013; Light and Cunningham 2016; McLevey 2021; Mohr et al. 2013; Nelson 2020; Roose et al. 2018). This type of model assumes that texts are composed of multiple topics that have different probabilities of occurring and that each topic is composed of words that have different probabilities of appearing. Bayesian inference is used to estimate the distributions of topics and words based on observed co-occurrences between pairs of words in the text corpus.

Usually, topic models are unsupervised, which means they are entirely *inductive* models that use bottom-up pattern detection methods to identify latent topics that researchers review and interpret. These types of topic models cannot be "steered" to focus on specific types of discourses. Therefore, we used a kind of *semi-supervised hierarchical topic model* called CorEx (Gallagher et al. 2017) that uses information theory rather than Bayesian inference to identify latent topics. This blends induction and deduction into one modelling framework and enables us to better link our computational analysis to findings from other parts of the project. Our approach is inspired by Laura Nelson's (2020) "computational grounded theory" framework.

We made several important decisions when developing topic models. We chose to develop models that would identify 23–30 topics. Selecting the appropriate number of topics is always a challenge, but there are guidelines that make the decision less arbitrary (McLevey 2021). In this case, we started by estimating 200 unsupervised topic models for each of the five cases, which initially produced 1000 topic models. We then looked for points at which adding more topics resulted in only minor changes to the total correlation explained. Across these 1000 models, we saw that models with topics in the 15–30 range were generally preferable to models with fewer than 10 topics, or more than 30 topics. From the initial models, we recorded a set of "anchor words" to use in successive rounds of model development.

The next part of our analysis involved iteratively improving the topic models for each of the five cases. We identified potentially important topics and keywords from the initial unsupervised models that clearly connected to findings from our qualitative analysis. At the same time, we identified keywords from our qualitative analysis that we wanted to "look for" in the news story data, such as "activist," "movement," and "protest." We continued this iterative process of refining anchor words based on *both* the qualitative findings and the computational induction of our topic modelling approach. To facilitate comparisons across all five cases, we used the list of anchor words as a baseline across all cases. However, we also added sets of anchor words for each case to better capture more case-specific topics (e.g. Scottish independence). We added 1–3 additional sets of anchor words for all cases except Denmark, where we found that adding new anchors words did not improve our models.

It is important to note that our goal was not to identify *all* potentially relevant words associated with some potential topic, as that would tip the scales too far on the side of deduction and impose too many constraints around what the model might find. We also did not attempt to exhaustively itemize every topic that could be relevant to this project. Instead, we added and removed terms and topics from this list as we worked across multiple datasets and compared preliminary results across cases.

Another important decision we made concerned how forcefully we steered the models towards finding topics that include some or all anchor terms. In this case, we developed the models with a light touch; if the model could not find meaningful topics with our anchor words, we did not force it. Instead, our models either disregarded the anchor term, or returned a "junk topic" made up of words that were weakly connected to the anchor terms. *These "junk topics" are findings, not indications of a bad model*. For example, if the anchor word "renewable" returns a topic that is clearly about renewable energies in one case but produces a junk topic when applied in another case, then we can conclude that renewable energies is a coherent topic in the former case but not in the latter.

For each of our five cases, we developed hierarchical topic models that cluster the results of the final baseline model for each case into a discourse network, where topics that are more closely related to one another are merged into more general discourse clusters. Topics are linked in a macro-level discourse network where the strength of the tie is determined by the degree of association between the two topics. We visualize the results of this hierarchical analysis in a series of figures in Chapter 3.

In addition to these semi-supervised topic models, we constructed sematic networks based on word co-occurrences within news stories. We then used community detection algorithms to identify potential topics within and across topics. The results of this analysis were compared with the results of the semi-supervised hierarchical topic models. While the latter provided more useful and informative results, the results of the two computational approaches overlapped considerably. That overlap gives us confidence that the topic model results are not an artefact of our anchoring strategy.

After initial coding and analysis in each of the methods described above, we engaged in within-case and cross-case comparisons and synthesis of data sources. Discourse network analysis (which differs from the semantic network approach to the media corpus) was one technique used to examine relationships among different themes, as well as relationships between themes and organizational actors (Leifeld 2017). Coding matrices were generated in NVIVO, which provided a view of the intersections of coding across multiple coding categories. These coding matrices were imported to Visone (software for social network analysis), where they were visualized as networks. This allowed us to generate 1-mode network diagrams that provide a higher-level view of the patterns of relationships among thematic categories or organizational actors. We also used 2-mode network diagrams that provide a higher-level view of relationships between organizational actors and thematic categories.

Qualitative comparison tables were another technique for data synthesis and comparison. Table 7.4 is a partial sample of the template used for each of the case study regions, which linked analytical dimensions (as rows) with the different data sources (as columns) for each case. As each within-case table was finished, it was shared for feedback with our project collaborators who have regional area expertise on Denmark, Iceland, Newfoundland and Labrador, Norway, and Scotland. This helped serve as a face validity check on our results.

Once the within-case tables were completed for each of the five case study regions, we synthesized the results using a comparable cross-case qualitative comparison table. The analytical dimensions (rows) were the same, with the case study regions as the columns, with a final column providing space for key points of comparison.

Table 7.4 Sample qualitative comparison table

Dimension	Corpus analysis (mass media)	Interviews	Field notes	Internet ethnography (netnography)	Document analysis	Key points of comparison
Culture: Mediated representations of coastal communities and environments as spaces for nature-based tourism and oil extraction						
Culture: Discourses of positive and negative social-environmental impacts of tourism development						
Culture: Discourses of positive and negative social-environmental impacts of oil extraction						
Culture: Dominant discourse of relationship between oil and tourism as different development pathways—antagonism, co-existence, or silos (black-boxing)						

Conclusion

This project weaves together multiple strategies for data collection and analysis to better understand the oil-tourism interface in sites across the North Atlantic. Our research process orients around methodological omnivorousness, methodological pragmatism, and mobile methodologies. This methodological approach to studying social-ecological relationships is useful for several reasons. First, there is a methodological emphasis on relationality that aligns well with the relational political ecology theoretical perspective that frames our analysis. Second, this approach offers a pragmatic and pluralist alternative to qualitative-quantitative dichotomies that persist in the social sciences, but which we see as unhelpful for creating complex (Law 2004) or "crystalline" (Richardson 2000) research accounts of social-ecological systems. Third, this approach allows us to examine the structure of players within arenas of environmental governance and collective action, as well as the cultural meanings at play in these arenas and processes of interaction. Fourth, and finally, this approach is well suited to analyses of cultural and socio-semantic networks, environmental governance, and social movement strategic interaction, across different social-political scales.

References

Alford, R. (1998). *The craft of inquiry: Methods, theories and evidence*. Oxford, UK: Oxford University Press.

Bail, C. (2016). Combining natural language processing and network analysis to examine how advocacy organizations stimulate conversation on social media. *Proceedings of the National Academy of Sciences, 113*(42), 11823–11828.

Blei, D. M. (2012). Probabilistic topic models. *Communications of the ACM, 55*(4), 77–84.

Blok, A. (2010). Mapping the super-whale: Towards a mobile ethnography of situated globalities. *Mobilities, 5*(4), 507–528.

Büscher, M., Coulton, P., Hemment, D., & Mogensen, P. H. (2011a). Mobile, experimental, public. In M. Büscher, J. Urry, & K. Witchger (Eds.), *Mobile methods* (pp. 119–137). New York: Routledge.
Büscher, M., Urry, J., & Witchger, K. (2011b). Introduction: Mobile methods. In M. Büscher, J. Urry, & K. Witchger (Eds.), *Mobile methods* (pp. 1–19). New York: Routledge.
Campbell, L. M., Corson, C., Gray, N. J., MacDonald, K. I., & Brosius, J. P. (2014). Studying global environmental meetings to understand global environmental governance: Collaborative event ethnography at the tenth conference of the parties to the Convention on Biological Diversity. *Global Environmental Politics, 14*(3), 1–20.
Collins, H. (2004). *Gravity's shadow: The search for gravitational waves*. Chicago, IL: University of Chicago Press.
Cresswell, J. W. (2003). Research design: Qualitative, quantitative, and mixed methods approaches (2nd Ed.), *Thousand Oaks*, CA: Sage.
DiMaggio, P., Nag, M., & Blei, D. (2013). Exploiting affinities between topic modeling and the sociological perspective on culture: Application to newspaper coverage of US government arts funding. *Poetics, 41*(6), 570–606.
Gallagher, R. J., Reing, K., Kale, D., & Ver Steeg, G. (2017). Anchored correlation explanation: Topic modeling with minimal domain knowledge. *Transactions of the Association for Computational Linguistics, 5*, 529–542.
Graham, A., McLevey, J., Browne, P., & Crick, T. (2020). *Structural holes and culturally constrained cognition: Clarifying the role of intellectual diversity and complementary information in scientific influence and impact* (Working paper).
Grimmer, J., & Stewart, B. (2013). Text as data: The promise and pitfalls of automatic content analysis methods for political texts. *Political Analysis, 21*(3), 267–297.
Hine, C. (2005). Internet research and the sociology of cyber-social-scientific knowledge. *The Information Society, 21*(4), 239–248.
Honnibal, M., & Montani, I. (2017). *Spacy 2: Natural language understanding with bloom embed-dings, convolutional neural networks and incremental parsing* (Working paper).
Ignatow, G., & Mihalcea, R. (2017). *An introduction to text mining: Research design, data collection, and analysis*. Thousand Oaks, CA: Sage.
James, L., Olsen, L. S., & Karlsdóttir, A. (2020). Sustainability and cruise tourism in the Arctic: Stakeholder perspectives from Ísafjörður, Iceland and Qaqortoq, Greenland. *Journal of Sustainable Tourism, 28*(9), 1425–1441.

Lamont, M. (2009). *How professors think: Inside the curious world of academic judgement*. Harvard, MA: Harvard University Press.
Law, J. (2004). *After method: Mess in social science research*. London: Routledge.
Leifeld, P. (2017). Discourse network analysis: Policy debates as dynamic networks. In J. N. Victor, M. N. Lubell, & A. H. Montgomery (Eds.), *The Oxford handbook of political networks* (pp. 301–326). Oxford, UK: Oxford University Press.
Light, R., & Cunningham, J. (2016). Oracles of peace: Topic modeling, cultural opportunity, and the nobel peace prize, 1902–2012. *Mobilization: An International Quarterly, 21*(1), 43–64.
McLevey, J. (2019). Epistemic and evidential cultures. In P. Atkinson, S. Delamont, R. Williams, & A. Cernat (Eds.), *Sage research methods foundations*. London: Sage. https://doi.org/10.4135/9781526421036840315.
McLevey, J. (2021). *Doing computational social science*. London: Sage.
Merriman, P. (2014). Rethinking mobile methods. *Mobilities, 9*(2), 167–187.
Mohr, J. W., Bogdanov, P., DiMaggio, P., Nag, M., Blei, D., McFarland, D., et al. (2013). Topic models and the cultural sciences. *Poetics, 41*(6), 545–770.
Morgan, D. L. (2007). Paradigms lost and pragmatism regained: Methodological implications of combining qualitative and quantitative methods. *Journal of Mixed Methods Research, 1*(1), 48–76.
Nelson, L. K. (2020). Computational grounded theory: A methodological framework. *Sociological Methods & Research, 49*(1), 3–42.
Nuttall, M. (2012). Imagining and governing the Greenlandic resource frontier. *The Polar Journal, 2*(1), 113–124.
Page, S. E. (2019). *The diversity bonus: How great teams pay off in the knowledge economy*. Princeton, NJ: Princeton University Press.
Ren, C., Bjørst, L. R., & Dredge, D. (2016). Composing Greenlandic tourism futures. In M. Mostafanezhad, R. Norum, E. J. Shelton, & A. Thompson-Carr (Eds.), *Political ecology of tourism: Community, power and the environment* (pp. 284–301). London: Routledge.
Richardson, L. (2000). New writing practices in qualitative research. *Sociology of Sport Journal, 17*(1), 5–20.
Roose, H., Roose, W., & Daenekindt, S. (2018). Trends in contemporary art discourse: Using topic models to analyze 25 years of professional art criticism. *Cultural Sociology, 12*(3), 303–324.
Rule, A., Cointet, J. P., & Bearman, P. (2015). Lexical shifts, substantive changes, and continuity in State of the Union discourse, 1790–2014. *Proceedings of the National Academy of Sciences, 112*(35), 10837–10844.

Scott, N. A. (2020). *Assembling moral mobilities: Cycling, cities, and the common good*. Lincoln, Nebraska: University of Nebraska Press.
Shadian, J. M. (2014). *The politics of Arctic sovereignty: Oil, ice and Inuit governance*. London: Routledge.
Small, J., & Harris, C. (2014). Crying babies on planes: Aeromobility and parenting. *Annals of Tourism Research, 48,* 27–41.
Small, M. (2011). How to conduct a mixed methods study: Recent trends in a rapidly growing literature. *Annual Review of Sociology, 37,* 57–86.
Small, M. L., & Adler, L. (2019). The role of space in the formation of social ties. *Annual Review of Sociology, 45*(1), 111–132.
Spinney, J. (2011). A chance to catch a breath: Using mobile video ethnography in cycling research. *Mobilities, 6*(2), 161–182.
Stokes, A. (2020). *Masters of (n)one: Negotiating portfolio careers and cultural legitimacy in creative industries* (Working paper).
Thorpe, H. (2014). *Transnational mobilities in action sport cultures.* New York: Palgrave Macmillan.

Index

M. C. J. Stoddart et al., *Industrial Development and Eco-Tourisms*, https://doi.org/10.1007/978-3-030-55944-1

B

C

L

M

N

O

The manufacturer's authorised representative in the EU is Springer Nature Customer Service Centre GmbH, Europaplatz 3, 69115 Heidelberg, Germany. If you have any concerns regarding our products, please contact ProductSafety@springernature.com

Printed and bound by CPI Group (UK) Ltd, Croydon, CR0 4YY
30/09/2026
02225451-0006